**The Open University**

*Science: a third level course*

# LIVING PROCESSES

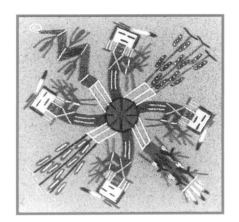 

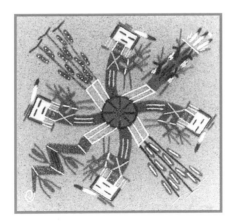 

## Book 3
## Biocommunication

*Edited by Steven Rose*

## The S327 Course Team

*Course Team Chair*
Steven Rose

*General Editor*
Mae-Wan Ho

*Academic Editors*
Anna Furth (Book 1)
Mae-Wan Ho (Book 2)
Steven Rose (Book 3)
Steve Jones (University College London) (Book 4)

*Authors*
Mary Archer (Imperial College) (Book 2)
Jim Barber (Imperial College) (Book 2)
Samantha Bevan (Book 3)
Brian Challis (Book 1)
Basiro Davey (Book 3)
Donald Edmonds (Oxford University) (Book 3)
Anna Furth (Book 1)
Catherine Gale (University College London) (Book 4)
Brian Goodwin (Book 3)
David Harris (Oxford University) (Book 2)
Roger Hill (Books 1 and 2)
Mae-Wan Ho (Book 2)
Tim Hubbard (MRC Cambridge) (Book 1)
Jim Iley (Book 2)
Steve Jones (University College London) (Book 4)
Judith Metcalfe (Book 4)
Steven Rose (Books 1 and 3)
Kay Taylor (University College London) (Book 4)
Jonathan Wolfe (University College London) (Book 4)

*Course Manager*
Jennie Simmons

*Editors*
Sheila Dunleavy
Kate Richenburg
Gillian Riley
Bina Sharma
Margaret Swithenby

*Design Group*
Diane Mole (Designer)
Pam Owen (Graphic Artist)

*BBC*
Rissa de la Paz
Phil Gauron

*External Course Assessor*
Professor R. J. P. Williams (Oxford University)

First published 1995

Edited, designed and typeset by The Open University.

Printed in the United Kingdom by Eyre & Spottiswoode Ltd, London and Margate

ISBN 0 7492 51557

This text forms part of an Open University Third Level Course. If you would like a copy of *Studying with The Open University*, please write to the Central Enquiry Service, PO Box 200, The Open University, Walton Hall, Milton Keynes, MK7 6YZ. If you have not enrolled on the Course and would like to buy this or other Open University material, please write to Open University Educational Enterprises Ltd, 12 Cofferidge Close, Stony Stratford, Milton Keynes, MK11 1BY, United Kingdom.

1.1

# Contents

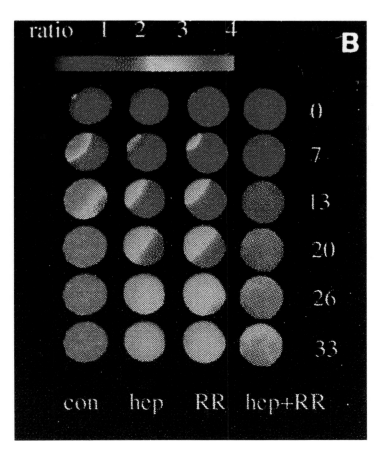

**Plate 1.1** Oscillations in cytosolic calcium levels in sea urchin eggs following fertilization. Calcium levels in the cytosol of the eggs were imaged by injecting a $Ca^{2+}$-sensitive dye into the eggs. The egg is 100 µm in diameter. High $Ca^{2+}$ concentrations are shown in orange, resting concentrations in blue, intermediate concentrations in yellow and green (see scale at top of picture). Successive images were recorded 6.6 s apart (see number scale in seconds on right). The fertilization wave begins at the site of sperm entry (10 o'clock position) and crosses the egg over a period of about 20 s (first column, control). In the second, third and fourth columns, the egg had been injected with inhibitors of egg activation prior to fertilization (heparin (hep), ruthenium red (RR) alone and hep + RR together).

**Plate 2.1** The electric eel *Electrophorus electricus*.

# *Introduction*

## 1.1 Organisms and environments

Book 1 of this Course focused on the chemical nature of a typical cell and the structure and properties of the molecules of which it is composed, especially the macromolecules. In Book 2 we moved from the somewhat static picture that this offered to consider the cell in action – how it generates and utilizes energy both to maintain its routine existence and to enable it to perform work on the outside world, as, for example, in muscle contraction. In Book 3, we consider how the multitudes of processes occurring at any one time within cells and within the organisms which they make up are controlled and regulated, and how order is sustained amidst such constant chemical flux. Essentially this is the problem described in Book 1, Chapter 1 as that of *homeostasis*, the maintenance of the internal environment of a cell within an acceptable range of temperature, humidity, pH, ion content, ATP concentration, etc. It is the problem approached again in Chapter 5 of Book 2, which introduced the idea of the metabolic web through which molecular transformations within the cell occur and are choreographed. But it is also more than this, because, again as we spelled out in Book 1, cells do more than simply maintain constancy – they are born, grow, age and divide and throughout their lives they interact with other cells of the same organism or with the external environment. Thus homeostasis has to be maintained against a constantly moving trajectory – a process we described as *homeodynamics*.

The problems this poses are different for unicellular and multicellular organisms. Unicells have very little capacity to control their immediate external environment. Floating in a fluid medium, an amoeba or bacterium may be faced with widely fluctuating levels of energy-providing substrates, ionic concentrations or pH. However, it is far from passive under such circumstances; to survive it has to be able to withstand such fluctuations, by adapting to changing conditions, and where necessary by migrating to a more favourable environment. If a bacterium is put into a test-tube of buffer solution and a drop of glucose solution added, the organism will use its cilia (which you saw in Video sequence 1) to move up the gradient of glucose concentration towards the energy source. It will feed on the glucose, absorbing and metabolizing it, and in due course excrete the waste products of its metabolism, thus in turn changing the environment. Some organisms are very specialized and can only survive in a very limited range of environmental conditions, but many unicells are highly adaptable. Glucose is not always the most available of foodstuffs and when it is absent a flexible organism must have the capacity to survive on other energy sources, the utilization of which may require a very different array of enzymes. It is thus in the nature of living organisms that they are not closed but open systems, actively choosing and modifying their environment, as well as themselves adapting to it; in short, living organisms are and must be in constant interaction with their environment.

However, 'the environment' is not merely the physical medium in which the organism finds itself; it includes other organisms too – conspecifics, predators and prey – to which even unicellular organisms need to be able to respond. Such responses may, depending on circumstances, be either competitive or cooperative. Individuals may compete for a limited food source or cooperate – for example in sex (even bacteria have a form of sex, called *conjugation*, in which they exchange genetic material). In times of hardship, such as limited food supplies or space in which to grow, populations of some single-celled organisms can dramatically change their life-style by coming together as a colony, in which the individual unicells combine in their millions to form a spore or fruiting body (one of the best known examples of this is the cellular slime mould, whose life cycle is shown in Figure 1.1). For such processes to occur, a form of communication must take place between the individual unicells. As you will see, such communication normally takes the form of a chemical signal which is produced by the individual cells, diffuses between them and changes their behaviour. Such chemical signals are better studied in multicellular organisms, where they are referred to as **pheromones**.

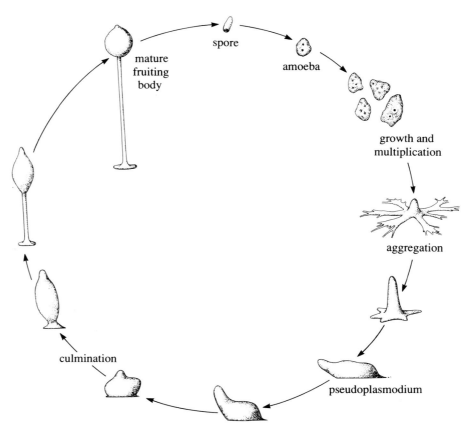

*Figure 1.1*  Life cycle of a slime mould.

Other forms of cooperation can take place between organisms of different species. Many live in a **symbiotic** relationship in which each depends for survival on the other (the nitrogen-fixing microorganisms which live in nodules on the roots of leguminous plants are a good example). In some cases the symbiosis has become so permanent a feature of their life-styles that the two organisms actually merge to become something qualitatively different from either, a process known as *symbiogenesis*. One well-established example of such symbiogenesis relates to the origins of mitochondria and chloroplasts, which, it is believed, were once free-living independent organisms that have, during evolution, become incorporated into animal and plant cells, losing their independence and many of their genes in the process – this example will be discussed in some detail in Book 4.

The problems faced by multicellular organisms are rather different to those of unicells; the individual cells of a human, a flatworm or an oak tree are not as independent entities. Instead they are arranged in tissues and organs, each functionally specialized and expressing a different subset of the many different genes the organism possesses. Sacrificing independence provides many benefits. For instance, except for the surface cells of the body, the 'environment' is no longer the external world with all its uncontrollable fluctuations but is *internal* to the organism, kept within homeostatic limits by the circulation of extracellular fluids, and in animals, the blood. Individual cells within a multicellular organism thus no longer require, and in most cases have lost, the adaptability shown by unicells to survive extremes, utilize wide ranges of foodstuffs, etc. It becomes the task of the organism as a whole to respond to environmental change whilst keeping its internal environment relatively constant. So, for instance, in humans mechanisms exist to maintain the internal body temperature within a degree or so of 38 °C whilst the external temperature may vary across a range of 50 °C or more, and to maintain blood glucose levels at around 5 mmol l$^{-1}$ in conditions ranging from satedness to acute starvation. But to achieve this internal constancy requires a far greater degree of **communication, coordination** and **control** between individual cells – the 'three Cs' – than is the case for the relatively autonomous unicells. Such coordination is achieved, as you will see in Chapters 4 and 5, through the development of highly specialized cells whose task within the organism is the generation and distribution of signals, and of detector systems on the surface of the target cells which can receive and respond to these signals.

Multicellularity thus radically changes the nature of the biocommunication processes. Whereas unicells have two levels of communication – one internal to each cell and the second with the external environment – multicellular organisms have three:

1   that internal to each cell;

2   that between cell and cell, and cell and internal environment within the organism;

3   that between the organism and its external environment.

In multicells, communication with the external environment (level 3) largely depends on the receptor and effector activity of specialized organs (e.g. sense organs, muscles) and is the province of physiology and neurobiology, falling outside the range of the present Course. The major theme of this book is that of the intermediate level (2), of communication at the interface between cell and cell, and cell and internal environment within an organism. However, in the final two chapters we do move outwards to consider aspects of the relationship of organisms with their external environment – including, as you will find at the end of the final chapter, their conspecifics. However, before we can discuss these mechanisms in detail it is necessary both to consider some general issues raised by the needs of biocommunication, and then to enter the interior of the cell to consider the intracellular processes that affect the 'three Cs'.

## 1.2   Processes of biocommunication

### 1.2.1   Space and time

Organisms can be regarded as a nested hierarchy of structures and processes: macromolecules embedded within organelles within cells within tissues within organisms. To each level of the hierarchy corresponds a particular set of dimensions in space and time. A typical human cell is some 5–10 μm in diameter; the extracellular gap between two communicating nerve cells is about 0.5 μm.

The person in whom these cells reside may be nearly 2 m tall. Thus the signalling processes which regulate and maintain order within the body have to span a spatial range of more than seven orders of magnitude.

Whilst this spatial hierarchy may be obvious, the temporal one is perhaps less so. A substrate molecule within the cell may be synthesized and broken down within a millisecond, a person may live for up to a hundred years, or $3 \times 10^9$ seconds, a difference of 12 orders of magnitude. Small molecules of substrates such as the intermediates of glycolysis or the tricarboxylic acid cycle are broken down and resynthesized thousands of times a second. It takes a few seconds to synthesize a new protein molecule, and if the protein is required to be transported to another part of the cell – to be inserted into a cell membrane, for instance – it can travel at the rate of some 16 mm an hour. Once synthesized, the average lifetime of any protein molecule in the body is around 10–15 days, but the range is huge, anything from a few minutes to many months. In the adult human, the average lifetime of cells too varies enormously depending on cell type; erythrocytes (red blood cells), for instance, are formed from precursor cells in the bone marrow, released into the bloodstream and once there survive for up to 120 days before they die and their contents are recycled. The longest-lived cells in the body are neurons (nerve cells) which, once born from their precursor cells (mainly during embryonic development), never divide and are never replaced by other neurons; the vast majority therefore survive as long as the life of the brain in which they are located (the space occupied by those that do die becomes filled either with nerve processes – axons and dendrites – from the remaining neurons or with a type of support cell called a glial cell). The number of cells of all kinds in the body is astronomic, especially during development when their rate of production is so rapid as almost to beggar imagination. For instance, it has been estimated that during the first few months of human life the rate of production of synapses – the junctions between nerve cells that you will learn more about in Chapter 4 – may be as much as 30 000 per second!

## 1.2.2 Electrical and chemical signalling

Such ranges in space and time vastly exceed the capacity one might expect of any current human technology or indeed any single biological signalling mechanism; as we shall see, cells and organisms utilize a wide range of different mechanisms adapted to the different levels of space and time over which communication is required. For something to act as a signal it must satisfy four criteria:

1    It must be rapid relative to the process which it is intended to influence.

2    It must be detectable by that process.

3    The process must be able to respond appropriately to it.

4    Once the signal has been detected it must be rapidly terminated.

Think of trying to let someone inside a house know that you are outside by ringing the doorbell. The bell must sound within a brief time after you have pressed it, but for much less time than it takes for the person inside to come to the door and let you in. If the ring is too faint, or the person inside is hard of hearing, it will not be detected and cannot serve as a signal. If the person inside doesn't recognize it as a doorbell but hears it instead as a burglar alarm, their response will be inappropriate. And if the bell goes on ringing long after the person inside has responded, not merely will it be a nuisance, but it will not be possible for later visitors to make themselves heard.

In the interior world of the cell or of the organism, signals are generally provided in the form of a localized change in concentration of a particular molecule or ion – for instance, the arrival at the surface of the cell of a peptide hormone, or a change in the distribution of ions across a membrane. It is conventional to distinguish between these two types of signal: the first form is described as chemical, the second electrical – although unlike the bell, which depends on the flow of electrons down a wire, electrical signals in living systems are based on the flow of cations across a membrane, thus altering the voltage across it. In general, biochemists have concentrated on study of the first type of signal, biophysicists and physiologists on the second. But as in each case the signal is provided by a change in local concentration or distribution of a molecule or ion, the distinction is less fundamental than this disciplinary specialization might imply.

As you already know in principle, and as you will learn in much greater detail in Chapter 2, the maintenance of a potential difference across the cell membrane is a fundamental feature of living systems, and the regulation of ion flux across the membrane is a vital homeostatic mechanism. It is the exploitation of this property by the specialized cells of the nervous system, the neurons, that provides the basis for the rapid signalling mechanism which multicellular animals have evolved, and which is discussed in Chapter 5. The more obviously biochemical signals, those provided by hormones that interact with receptors on or in the cell membrane, and whose presence results in the generation of a second set of chemical signals inside the cell (second messengers), are the subject of Chapters 3 and 4.

A key feature of living systems is their need to survive against the background of an environment which itself is rich in other life forms. Whilst we have emphasized the extent to which organism and environment interpenetrate, it is still necessary for self-preservation for any individual organism to be able to maintain its boundaries, to protect itself from invasion and occupancy by other organisms which, far from relating to it symbiotically, themselves survive only by weakening or destroying their host. To protect itself in this way an organism needs the capacity to distinguish *self* from *non-self*, so as to be able to recognize, react to and if necessary inactivate, destroy or expel such unwanted invaders within its boundaries. This is the function of the body's immune system, an elaborate array of specialized cells and proteins that have evolved to perform such recognition functions. The system is so complex that it could readily warrant a book in itself; here the bare bones of how the system works as a communication mechanism will be found in Chapter 4. Through each chapter you will find yourself moving steadily outward from the individual cell, a process which continues in Chapter 6, which discusses the role of electric and magnetic currents and fields at the level of the organism itself and its interactions with the external world. Finally, Chapter 7 and Case Study 3 consider the ways in which oscillatory processes at a multitude of levels contribute to homeodynamic stability and the maintenance of order.

### 1.2.3 Stability through dynamic change

The maintenance of stability is, as we have constantly emphasized, fundamental to the preservation and development of living systems. Paradoxically, however, such stability is best preserved in a dynamic rather than a static system. To see why, consider how temperature is maintained in a house by the thermostatic control in a central heating system. If one sets the control for a specific temperature, say 20 °C, and then measures the actual room temperature, what one finds is not that temperature is maintained at a steady 20 °C, but that there are cyclical fluctuations around that temperature, as shown in Figure 1.2.

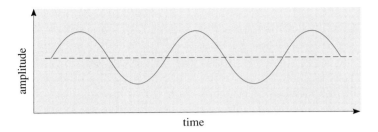

**Figure 1.2** Cyclical fluctuations around a set value (dashed line).

The reasons for this are obvious; the thermostat has a sensitivity range equivalent to 1–2 °C. As the central heating system comes on the room warms up to say 21 °C, triggering the thermostat to switch off; the room then cools to 19 °C, and the thermostat switches on again. If the thermostat were too sensitive, say to only 0.1 or 0.01 °C, then the heating system would switch itself on and off with extreme rapidity, which would both be inefficient and unlikely to preserve the life of the equipment. In designing a system to maintain homeostasis, therefore, a balance has to be struck between the permitted range of deviation from the desired steady state and the sensitivity of the homeostatic mechanisms. Thus stability is provided not by struggling to maintain absolute constancy but by encouraging rhythmic oscillations around an appropriate mean.

What is true for human artefacts is also true for biological systems such as cells. But this is only the first level of homeostatic control, for cells are embedded in organs whose physiological functions may require them too to show oscillations. Everyone is familiar with the rhythmic processes of breathing or heart beat, which are discussed in detail in Chapter 7. These rhythms too are embedded within larger cycles. Nearly all multicellular organisms show **circadian rhythms**, oscillations in behaviour with a periodicity of about 24 hours, of which the most obvious in humans, though far from the only one, is the cycle of sleep and wakefulness, a cycle which is expressed at many levels, from the biochemical through the physiological to perhaps the most immediately obvious, the behavioural. Over and above these daily rhythms are monthly ones (such as the menstrual cycle) and even yearly ones. These too will be discussed in Chapter 7. Thus one could amplify the simple diagram of Figure 1.2 to show the superoscillations of Figure 1.3.

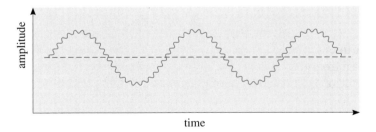

**Figure 1.3** Oscillations and superoscillations in a biological system.

But even this is too simple, for it concentrates on the maintenance of stability – stasis – at the expense of the dynamic change which characterizes living organisms through their life cycle. A more truly homeodynamic representation would superimpose these nested homeostatic cycles on the trajectory given by birth, development, ageing and ultimately death (see Figure 1.4). Even this representation speaks only of the dynamics of a single organism; populations and communities of organisms, and species through evolutionary time, all show such homeodynamic oscillations, which, though fascinating in their own right, are beyond the scope of this Course – although we shall be looking briefly at some of these phenomena and processes towards the end of Chapter 7.

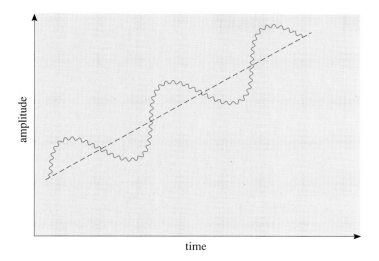

**Figure 1.4** Homeodynamic oscillations. In this very simplistic representation, the magnitude of the oscillating factor increases steadily with time, while the magnitude of the superoscillations remains constant.

## 1.3 Regulation and control within the cell

The story of biocommunication related in the subsequent chapters of this book begins within the cell and moves outwards towards the organism and even populations of organisms. You will see how, as this journey outwards is made, so space and time perspectives shift, from fast short-range forces and processes to longer-lasting and longer-range ones. But for the rest of this chapter and, as a way of providing the base from which the rest of the book can build, we shall return to the cell itself, and concentrate on the general principles by which intracellular order is maintained.

As you know from Books 1 and 2, the internal milieu of the cell is a structured mix of up to 30 000 different proteins, together with DNA and RNA. Some of these macromolecules are embedded within the phospholipid membranes of organelles, others are present in the fluid matrix of organelles and still others are dissolved in the cytosol. Within this inhomogeneous environment small molecules are continuously synthesized and degraded and concentrations of ions and cofactors fluctuate. Yet despite such constant flux the cell is in a state of relative stability. What are the factors that contribute to this stability? This was a question asked, and partly answered, in Chapter 5 of Book 2, which introduced you to the idea of the metabolic web, the distributed control of metabolism, and to microcompartmentation within an individual cell or organelle. Here, to begin with, we return from that degree of complexity to the factors that need to be taken into consideration in the context of control of a single reaction or reaction pathway.

First, consider a single reaction carried out in isolation in a test-tube, say the conversion of substrate X to product Y by the enzyme X-ase:

$$X \xrightarrow{\text{X-ase}} Y$$

▷  What are the factors that affect the rate and direction of this reaction?

▶  They will include the concentrations of X and Y, the thermodynamic equilibrium of the reaction, the amount of enzyme present (a small amount of enzyme will soon become saturated with substrate and will become the sole limiting factor), temperature, pH, ionic strength and the presence of any required ions or cofactors.

Any or all of these factors will affect the rate of reaction, that is, the number of molecules of X converted into Y in a given time, but in general it is likely that in any particular conditions one variable will have the most significant effect. For instance, small changes in temperature or pH may be of negligible significance and all reactants may be present in optimal concentrations but the enzyme may be obligatorily dependent on the presence of magnesium ions. Under such circumstances very slight alterations in the amount of magnesium present may have a very large effect on the rate of production of Y.

### 1.3.1 Feedback and rate-limiting steps

Now consider the variables affecting a reaction pathway, say the conversion of W to Z by way of the intermediates X and Y:

$$W \underset{\text{W-ase}}{\rightleftharpoons} X \underset{\text{X-ase}}{\rightleftharpoons} Y \underset{\text{Y-ase}}{\rightleftharpoons} Z$$

For each of the three reactions involved, the set of variables described above will control the reaction rate. But, over the entire sequence, an additional factor will now operate, one that does not exist for an isolated reaction. For a sequence of reactions the rate of each reaction is also affected by the outcome of the others. Suppose, for example, that the W-ase reaction is extremely sensitive to changes in pH, and that the final reaction product, Z, is an acidic substance. As more and more Z is produced, the acidity of the solution will increase. But when this happens, the enzyme W-ase will be inhibited and the rate of production of X from W will therefore be slowed. As the amount of X produced declines, so will the amounts of Y and Z. Thus the Z concentration will cease to rise and the acidity of the solution will decrease. Immediately, the rate of W-ase will once more be accelerated, so Z concentration and acidity will rise, and W-ase activity will therefore decline. Ultimately, a steady state will be reached.

Thus the production of Z exerts a controlling influence on W-ase and, reciprocally, the rate of W-ase controls the production of Z. This type of control is called **feedback**. It is identical in type to the principle of a the room thermostat described in the previous section. We can represent a reaction sequence with feedback like this:

$$W \underset{\text{W-ase}}{\rightleftharpoons} X \underset{\text{X-ase}}{\rightleftharpoons} Y \underset{\text{Y-ase}}{\rightleftharpoons} Z + (H^+)$$

Feedback can be of two types. In the example above, Z production inhibited W-ase – this is **negative feedback**. But if Z production lowered the pH so that W-ase was accelerated and Z production increased, this would be **positive feedback**. In all feedback reactions, the principle is the same – the rate of a reaction is controlled by a substance which is an ultimate product but not itself directly involved in the reaction.

When we come to consider the thousands of reactions and hundreds of reaction sequences that occur within the cell, all of them controlled by many different variables, and many themselves altering these variables by generating acidity or alkalinity, using or producing cofactors and inhibitors, our immediate feeling may well be one of despair. With so much going on, how can one sort out just what is, and what is not, significant?

But though it certainly remains complex, one can simplify the problem. Consider again the reaction sequence

$$W \underset{\text{W-ase}}{\rightleftharpoons} X \underset{\text{X-ase}}{\rightleftharpoons} Y \underset{\text{Y-ase}}{\rightleftharpoons} Z$$

If we ask what is the maximum rate, under the most favourable circumstances, of each of the enzymic reactions involved, and then express these rates in arbitrary units, we may get the results shown in Table 1.1.

What this table shows is that the W-ase reaction is ten times as fast as the X-ase reaction, and one hundred times as fast as the Y-ase reaction.

▷ What is now the **rate-limiting step** in the production of Z from W?

▶ Clearly W-ase is producing X at 10 times the rate that X-ase can use it, whilst X-ase is also producing Y at 10 times the rate that Y-ase can use it. As both X and Y are thus being produced far in excess of the amounts that can be handled by their respective enzymes, the rate of conversion of W into Z is limited only by the rate of the Y-ase reaction, $Y \rightleftharpoons Z$.

*Table 1.1*

| Reaction | Rate (arbitrary units) |
|---|---|
| W $\rightleftharpoons$ X | 100 |
| X $\rightleftharpoons$ Y | 10 |
| Y $\rightleftharpoons$ Z | 1 |

As a general rule, the rate of a reaction sequence is controlled by the rate of the slowest of the individual reactions of that sequence. Under these circumstances, even if Z production results in a change in pH big enough to cause a 10-fold decline in the rate of W-ase, the rate of Z production will remain unaltered, as this depends not on W-ase but on Y-ase. Z will no longer be exerting feedback control over its own production.

Thus the only reaction that matters as far as the control of a reaction sequence is concerned is the slowest reaction of the sequence, and the rate of any given sequence of reactions is regulated by altering the variables that control the rate of just one of those reactions – the slowest. This is where a study of enzyme kinetics and a knowledge of parameters such as $v_{max}$, as described in Book 2, can be helpful. You may thus expect to find in every reaction sequence one or more critical control points and you may also expect to find them at an early reaction in the pathway, as this would prevent the futile use of enzymes further along the pathway and the build-up of useless intermediates. Similarly, you could expect control points to occur very close to the branch points of pathways that serve more than one function, so that one branch could be regulated without affecting the others.

### 1.3.2  Calcium and phosphorylation cascades

When actual examples of metabolic pathways and their control are studied, these are indeed the principles which one finds employed. The enzyme that controls the rate-limiting step in a reaction sequence, generally the one towards the beginning of the particular sequence, often also turns out to be very sensitive to changes in concentration of cofactors, reactants or products, pH, temperature, etc. As you will discover in Chapter 3, one of the key factors is the concentration of calcium ions, and a rapid change in $Ca^{2+}$ concentration, perhaps brought about by a hormone interacting with its receptor on the external cell membrane, is often the signal for the initiation of a reaction sequence. In recent years, methods have been developed for measuring intracellular calcium concentrations, based on the injection into the cell of a dye which fluoresces in the presence of calcium. The use of such methods has shown that in response to such external signals rapid oscillations of cytosolic calcium levels can occur (Plate 1.1). You will learn more about the significance of these calcium signals in Chapter 3 and 4.

Another extremely important mechanism by which control can be achieved over the first step in a reaction sequence is to alter enzyme activity by modifying the enzyme itself. As you will recall from Book 1, several of the amino acid residues within a protein chain (e.g. serine, threonine and tyrosine) can be reversibly phosphorylated, usually at the expense of ATP. Phosphorylating the enzyme can result in a conformational change which alters – generally increases – its activity. This process is itself achieved enzymically. A class of enzymes called protein **phosphokinases** phosphorylate enzymes; a second class, phosphoprotein **phosphatases**, in turn dephosphorylate them.

Thus, control over the rate-limiting step in a reaction sequence can be achieved by a second reaction sequence which alternately phosphorylates and dephosphorylates the key enzyme, as shown in Figure 1.5 for the enzyme X-ase.

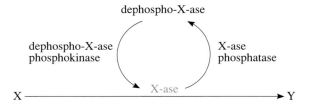

**Figure 1.5** Phosphorylation and dephosphorylation of a rate-limiting enzyme.

Some phosphokinases (sometimes called just kinases for short) are themselves activated by calcium ions, others by cyclic AMP. But to make matters more complex, and the level of control still finer, some phosphokinases are *themselves* activated or inhibited by reversible phosphorylation and dephosphorylation. Thus, the control points of important reaction sequences can be fine-tuned by a **cascade** of reactions, as shown diagrammatically in Figure 1.6. One way to think

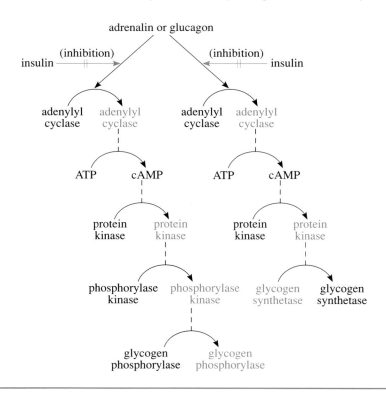

**Figure 1.6** Phosphorylation cascades involved in the control of glycogen catabolism. (The active forms of the enzymes involved are shown in colour.)

14

about such cascades (which have formed one of the most active areas of biochemical research over the past decade) is by analogy with the steps by which a coal fire is lit – you use a match to ignite paper, which ignites kindling, which in turn ignites the coal. You will come across many examples of such cascades in Chapters 3 and 4.

### Summary of Section 1.3

The rate of any individual reaction is affected by a variety of factors including temperature, pH, concentration of reactants, products, cofactors, etc. However, the rate of a reaction pathway is determined by the slowest reaction in that pathway. This reaction often occurs at the beginning of the pathway and may itself be subject to feedback control. A fine level of control is also exerted by regulating the activity of the enzyme itself; such regulation is achieved by covalent modification of the enzyme through phosphorylation and dephosphorylation. In many cases cascades of regulatory processes occur.

## 1.4 Control of biosynthetic pathways

Much of Book 2 was concerned with the cellular processes involved in the generation of utilizable energy in the form of ATP, notably by way of photosynthesis and glucose oxidation. One major example of the utilization of that energy – in muscular contraction – was also provided. But of course much of the cell's ATP is required for the maintenance of its own internal structures, even before that cell can begin to act on the external world by hormone secretion, nerve impulse or muscle contraction. Macromolecules, which are, as you know, in a constant state of flux, have to be synthesized from their component sugars, amino acids and fatty acids. And even these building blocks may themselves require synthesis if they are unavailable through food intake. Such synthetic reactions require energy, and each pathway involves at least as complex a series of enzyme-catalysed steps as that documented for glucose catabolism in Chapter 4 of Book 2. In this Course, there is neither enough time nor space to consider any of these synthetic pathways in biochemical detail. However, in the context of the 'three Cs' theme of this book, it is worth spending a little time considering some general mechanisms by which such biosynthetic pathways are controlled, and how they resemble and differ from the controls on energy production presented in Book 2.

### 1.4.1 End-product inhibition

Each biosynthetic pathway requires a chain of enzymes to build complex molecules from simpler molecules. Each chain has its own specific rate-limiting steps and feedback mechanisms, and many of these have now been analysed in detail, particularly in bacteria which are more biosynthetically versatile than animals.

Although each biosynthetic system shows certain unique features, there are some general mechanisms which apply. One such was provided by Edwin Umbarger, in New York, in 1956. He was studying the biosynthesis in bacteria of the amino acids threonine, lysine and methionine from glucose by way of a simpler amino acid, aspartate, and its metabolite aspartate semialdehyde (Figure 1.7). When the bacteria were fed radioactive glucose, the radioactivity was subsequently found in all three amino acids. However, when the organisms were grown in a medium containing, for instance, unlabelled threonine, the incorporation of radioactivity into this particular amino acid was abolished.

▷ How can this effect be accounted for?

▶ It must mean that the presence of the amino acid results in a suppression of its continued synthesis – a fine example of negative feedback.

This process, called, as you may recall from Book 2, Chapter 5, **end-product inhibition**, means that the biosynthesis of a given amino acid is regulated by the amount of the amino acid itself being produced as the end-product of a reaction sequence. As this amount increases, so further biosynthesis is reduced; if its concentration falls, synthesis starts up again. This form of feedback control has since been found to be widespread in biosynthetic systems, in animals as well as microorganisms, and is probably a major regulatory mechanism. Its most interesting feature can be illustrated if the pathway is drawn as in Figure 1.7.

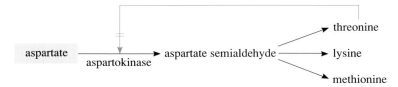

*Figure 1.7* Biosynthesis of amino acids from aspartate showing end-product inhibition of the first enzyme in the sequence.

Umbarger was able to show that the observed suppression of threonine synthesis was caused by the threonine inhibiting the first enzyme along its biosynthetic pathway, i.e. aspartokinase. However, production of the other amino acids was unaffected, even though they too are made from the same precursor, aspartate semialdehyde. This is because there is not simply one aspartokinase enzyme but three different **isoenzymes** each with slightly different structures, each specifically inhibited allosterically by one of the three amino acids, threonine, lysine or methionine. All three must be inhibited before production of the intermediate ceases completely. In addition, each of the amino acids inhibits the first enzyme on its branch line beyond aspartate semialdehyde (not shown in Figure 1.7), so ensuring that the decrease in concentration of this metabolite affects only the production of the inhibiting amino acid. The existence of multiple isoenzyme forms (**enzyme multiplicity**) thus affords a powerful regulatory mechanism.

A control mechanism that achieves the same effect as this, although by a different mechanism, is the **multi-end-product inhibition** of allosteric enzymes. Glutamine synthetase (responsible for the production of glutamine from glutamate) is regulated by this mechanism. It has binding sites for trytophan, histidine, CTP (cytidine triphosphate) and ATP, among others; this is because glutamine acts as an amino group donor in the biosynthesis of all these compounds. If only one of these binds to the synthetase then glutamine production will be turned down by a small amount. If two of them bind, then production will be lowered a little more. And so on. The effect is thus progressive, until all the end-products have been bound, when glutamine production ceases altogether. (In addition, as before, each of the end-products inhibits the first enzyme on its branch line from glutamine.)

### 1.4.2 The operon

An even more ingenious method of control was first identified amongst microorganisms by Jacques Monod and Francois Jacob in Paris, work which won them the 1965 Nobel Prize for Physiology or Medicine. They showed that many microorganisms that normally lacked the enzymes to deal with particular substrates, would, if grown in a medium containing these substrates, at once begin to synthesize the missing enzymes. In particular, Monod and Jacob studied bacteria lacking the ability to metabolize the sugar galactose. Presented with galactose as substrate, the bacteria of one strain proceeded to fabricate the enzymes needed to utilize it. The new enzymes that were required were synthesized completely from scratch – that is, a sequence of novel protein synthesis was initiated. This enzyme synthesis could be triggered even if galactose itself was replaced by a non-metabolizable substance which resembled it closely in structure (i.e. a *substrate analogue*). As the enzymes began to be made immediately in response to the substrate, the genetic information necessary for their synthesis must have been present in the cells all along. But the proteins were not normally present. How could this intriguing effect be explained?

Monod and Jacob argued that it occurred because, in the absence of the unusual substrate, the synthesis of the enzymes required to metabolize it was **repressed**, thus ensuring that the cell did not waste time or energy fabricating enzymes for a biosynthetic pathway it would not require. In the presence of the substrate, enzyme synthesis was **induced** by the added galactose or galactose analogue.

If the mRNA coding for the new enzymes is already present, then induction of their synthesis can occur by stimulating its translation. If it is not present, however, it would be necessary to make new mRNA molecules by transcription from the corresponding DNA. Bacterial mRNA, unlike the mRNA of animals, has a very short life, half of it decaying within about three minutes, so it seems unlikely that synthesis could be controlled by regulating translation. If new mRNA molecules were made, however, then transcription of the new mRNA could continue as long as the inducer was present. In the bacteria studied by Monod and Jacob, the genes controlling the synthesis of all the enzymes required for galactose metabolism (their **structural genes**) are situated along a consecutive length of DNA, and a little way off is another gene responsible for the manufacture of a specific repressor protein. As you know, mRNA is transcribed by a polymerase enzyme that binds first of all to a DNA **promoter** site. In the absence of the inducer, the repressor protein, according to the Jacob and Monod hypothesis, binds to a site on the DNA between the promoter and the structural genes and thus prevents the RNA polymerase from moving along from the promoter to transcribe those structural genes. The site to which the repressor protein attaches is called the **operator** region and the whole system of structural genes, repressor gene, and operator region is called an **operon**. The inducer exerts

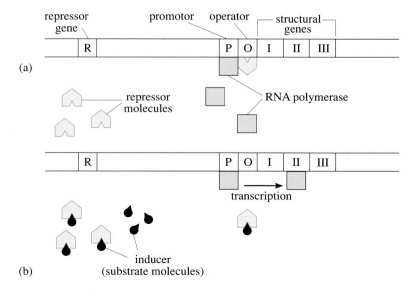

*Figure 1.8* The operon. (a) Repressor is bound to operator so transcription is prevented. (b) Inducer (e.g. substrate or substrate analogue) binds to repressor, so repressor is released from operator and transcription can proceed.

its effect by specifically combining with the repressor molecules so that they can no longer bind to the operator. RNA polymerase thus becomes unobstructed and so transcription can now occur.

A cell that can control its protein synthesis so as to make only the enzymes immediately demanded by its metabolic situation – even though it still retains the genes coding for many other enzymes – is clearly highly adaptable and well qualified to survive extreme variations in conditions, e.g. type of substrate available. It is not surprising therefore that regulatory systems based on the operon principle have been found to be widespread among micro-organisms.

### 1.4.3 Transcription factors

The whole system of protein synthesis is much more complicated in multicellular organisms. Unlike bacterial DNA, mammalian DNA is associated with the histones and the nucleohistone complex prevents transcriptional access to much of the DNA most of the time. This seems to be due to the histones stabilizing the DNA in such a way as to makes chain separation by breaking of hydrogen bonds more difficult. In addition, most of the RNA synthesized in the nuclei of multicellular organisms is subsequently degraded before it ever enters the cytosol, thus providing a further source of control. The initial product of transcription, the so-called primary mRNA transcript, is subject to a great deal of further processing in the nucleus before being permitted to leave by way of the nuclear pores. Sections of the primary transcript corresponding to particular exons are spliced together to give the mRNA molecules that leave the nucleus to serve as the templates for translation in the cytosol. For the reasons described in Section 1.1, the need for sensitive regulation of mRNA production is very different in multicellular organisms than in bacteria.

▷ Can you recall the key reason?

▶ Individual cells in a multicellular organism are buffered by the internal environment of the organism so do not normally need to adapt to such dramatic changes in the external environment.

However, the cells of multicellular organisms do need to be able to recognize and respond to the requirements of the organism as a whole if the organism itself is to function in an integrated manner. As you will see in Chapters 3 and 4, such regulation and integration is provided by the organism's nervous, hormonal and immune systems, and by the presence of elaborate receptor systems within the external cell membrane which can receive and respond to signals provided by these systems. In addition, the pattern of proteins expressed by individual cells within a multicellular organism needs regulation, especially during development. There are some 100 000 or so genes coding for proteins in the human genome, but most cells express no more than 10 000 or so different proteins. The cells that express the largest number of different proteins are the neurons which may have up to 30 000. Which proteins are expressed in a particular cell depends both on the tissue type (many of the proteins of liver cells are different from those of muscle cells, for instance) and on the developmental stage of the organism – fetal cells express different proteins from mature ones for instance. Thus, whereas bacteria may need to change their pattern of expression from moment to moment in response to availability of substrate, in multicellular organisms the pattern of gene expression must be **developmentally regulated**. To discuss the intimate details of such regulatory processes will take us further afield than is appropriate here, although we will return to the theme in Book 4. The important point to make here is that the process of developmental regulation, of turning particular genes on and off in orderly sequence, is a major method of control within the cell, and depends upon a class of proteins called **transcription factors** (TFs). These are themselves the products of genes known as *immediate early genes* and provide the intracellular signals whereby particular genes (*late genes*) become switched on and the proteins relevant to the tissue type and developmental stage of the cell are synthesized. The control features of this process are summarized in Figure 1.9.

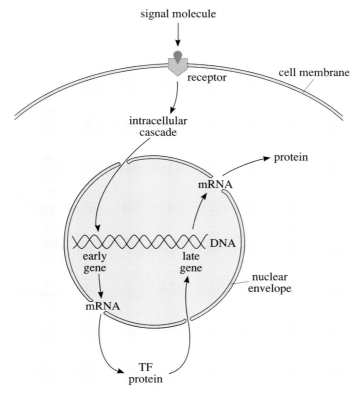

**Figure 1.9** Control by transcription factors.

## Summary of Section 1.4

Unicellular organisms, which rely on fluctuating sources of food, utilize end-product inhibition to prevent overproduction of abundant substances. In the absence of particular substrates, the synthesis of the enzymes which metabolize them is also often switched off; when the substrate again becomes available, the necessary enzymes can be induced. Multicellular organisms make less use of such mechanisms, in part because their individual cells exist within a more regulated extracellular environment. However, the production of particular proteins is regulated through controls on the transcription (and translation) of DNA. A special class of genes, immediate early genes, and their protein products intervene to regulate these transcription processes.

## 1.5   Structural constraints

The control processes described in the previous sections have been discussed as though the reaction sequences concerned were taking place relatively unconstrained within a cell protected from the external world by its cell membrane, but with its internal constituents in relatively free communication. This is, of course, as we were at pains to spell out in Book 2, a fiction. The cell is compartmentalized just as the organism is. Each internal organelle has a specialized function and a particular complement of enzymes and reaction sequences, and communication between organelles depends on the controlled transport of ions and metabolites across their membranes. For example, you have seen in Chapter 4 of Book 2 how the enzymes of the tricarboxylic acid cycle are sequestered in the mitochondrial matrix and the carriers of the electron transport chain embedded within the surrounding inner mitochondrial membrane. You will also know from the Science Foundation Course – and we have referred to it again in the previous section – how in eukaryotes the cell's DNA- and RNA-synthesizing machinery is contained within the nucleus, so that controlling the exit of mRNA through nuclear pores into the cytosol becomes an important further level of regulation of protein synthesis.

In the following chapters you will learn about the role of the external cell membrane in maintaining ionic and potential differences between the inside and the outside of the cell, and the function which this process serves in regulating cellular homeostasis. Examples of such control and signalling mechanisms can also be seen between the mitochondrion and the cytosol. Mitochondria contain substantial stores of calcium ions and a powerful active transport mechanism, analogous to those you will learn about in later chapters, enabling them to take up calcium from the cytosol against a concentration gradient. As we said in the last section, transient changes in calcium concentrations in the cytosol serve as signals for a multitude of enzyme cascades; thus, release of calcium from sequestration within the mitochondrion, as well as as entry of calcium from outside the cell, can provide such signals.

Within the cell the movement of newly synthesized proteins, and of materials intended for export, occurs through the interior of the membranous tubules of the endoplasmic reticulum and the Golgi apparatus, thus isolating and protecting these products from the rest of the cell; other substances, such as the neurotransmitters, are packaged in vesicles and moved via the microtubules away from the cell body where they are synthesized, to the regions of the cell where they are required (Figure 1.10).

But one of the most dramatic examples of control by structure within the cell is provided by the lysosomes (if you need reminding about these, see Figure 1.3 in Book 1, Chapter 1). These membrane-bound organelles contain hydrolytic enzymes (*hydrolases*) – proteinases, phosphatases and others – that is, enzymes capable of destroying all the macromolecules within the cell if they were to be free in the cytosol. In healthy cells, these enzymes – most of which have an optimum activity at acid pH – are retained within the lysosome, and so the cell remains safe. Instead, lysosomes entrap and engulf foreign material, which could be harmful to the cell, and digest it. You will learn more about this process in Chapters 3 and 4. However, in the event of disease or damage to the cell, the intracellular pH falls, the lysosomal membranes rupture and the hydrolases are released into the cell, rapidly digesting its contents and making them available for recycling or excretion. In this, the lysosome serves as a sort of suicide capsule for the cell, destroying it in the overall interest of the organism of which it is part.

This segregation of necessary but potentially lethal enzymes is but the most extreme example of the ways in which cellular homeostasis is maintained not merely by the checks and balances of enzyme activation, rate-limiting steps, feedback control and regulatory cascades, but by the intricate internal organization of each of the million million cells in each human body. These internal processes provide the necessary background against which to consider the increasingly higher levels of control, coordination and communication which the following chapters describe.

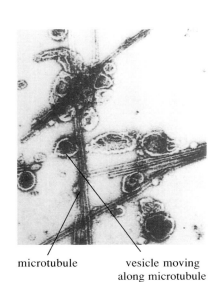

microtubule       vesicle moving
                  along microtubule

*Figure 1.10*  Transport of vesicles along microtubules.

## Objectives for Chapter 1

After completing this chapter you should be able to:

**1.1**   Define and use, or recognize definitions and applications of, each of the terms printed in **bold** in the text.

**1.2**   Define the three levels of environment within which a multicellular organism must operate.

**1.3**   Explain why rhythmicity imparts stability.

**1.4**   Describe the principles of control and the role of regulatory cascades within a metabolic pathway.

**1.5**   Contrast mechanisms of control of biosynthesis in unicells with those more prevalent in multicellular organisms.

## Questions for Chapter 1

### Question 1.1   (Objective 1.2)

Why is the meaning of the term 'environment' quite different for a liver cell and an *E. coli* bacterium in the gut?

### Question 1.2   (Objective 1.3)

Why would it be a mistake to design a rapidly responding room thermostat sensitive to 0.1 °C?

### Question 1.3   (Objective 1.4)

Consider the hypothetical metabolic pathway shown in Figure 1.11.

(a)  Which would you expect to be the rate-limiting step(s) in the pathway?

(b)  Describe and draw the pathway showing how end-product inhibition might work.

(c)  How might the enzyme A-ase be reversibly covalently modified so as to modulate its activity?

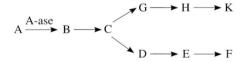

**Figure 1.11**   For Question 1.3.

### Question 1.4   (Objective 1.5)

Name one key system for the control of biosynthesis in unicells which plays a much smaller role in multicells.

## CHAPTER 2

*Intracellular electrical signalling*

## 2.1 Introduction

As we described in Chapter 1, communication within an individual cell can be by either chemical or ionic signals. Signals transmitted by the movement of charged ions are best analysed in terms, not of chemistry, but of electricity. This chapter explains why this is the case, and introduces you to the electrical properties of cells and their membranes. Electrical effects are of relatively minor importance at the level of the whole organism, but on the microscale of the cell and its individual subcompartments, they become of greater and greater significance. On the other hand, forces that are of importance to organisms the size of humans or oak trees, such as gravitation, become largely irrelevant at the microscale. This is because at the atomic level the only forces that exist outside the nucleus are electromagnetic or gravitational and on this size scale gravitational forces are too weak to have a significant influence. All other interactions such as mechanical force or chemical binding have at their core electrical interaction.

Electrical signalling within a living system is often achieved by the manipulation of ionic concentrations across membrane boundaries and a discussion of how this is achieved forms a large part of this chapter. Longer-range electrical communication between separated cells requires special mechanisms, and we shall defer consideration of this for a couple of chapters, in order to discuss chemical signalling, before returning to the electrical theme in Chapter 5.

*In working through this chapter, you will come across a number of biophysical formulae, equations and derivations. You are* **not** *expected to be able to remember all of these. As you will see from the objectives for this chapter, you are expected to learn and understand the derivation of one specific formula, that for the Nernst potential. For some of the other equations, you are expected to be able to interpret the formulae in words and to use them for specific worked examples, but not necessarily to be able to derive them from the first principles enunciated in the text (for guidance, see objectives, in-text questions and numbered questions).*

## 2.2 Ions and electricity

### 2.2.1 Ionic conduction

Our normal everyday experience of electricity is the domestic electricity supply. This is based upon metallic conduction in which the electric current is carried by electrons which are extremely light and mobile. As a result the speed of signal transmission along a metallic wire can approach the speed of light. In a living cell, however, the current carriers are not electrons, but ions in aqueous solution, which move much more slowly by a process known as **diffusion**.

An ion in solution is constantly colliding with neighbouring water molecules and each time it collides it moves very slightly under the impact. As an example, consider a potassium ion ($K^+$) which has a radius of 0.133 nm and the charge of a proton ($+e = 1.6 \times 10^{-19}$ coulomb). In water at room temperature such an ion is struck by a neighbouring water molecule about $10^{10}$ times each second and each collision results in a small displacement of the ion, much less than its radius. The motion that results from these rapid but random and directionless small displacements is called a **random walk**.

If a particle moves in a straight line with a constant speed, the distance travelled grows in direct proportion to the elapsed time. However, in a random walk such as those sketched in Figure 2.1, there is no preferred direction for each step and the average distance travelled from the starting point ($r$) grows only with the square root of the elapsed time $t$. The equation connecting $r$ and $t$ is

$$r = \sqrt{(6Dt)} \tag{2.1}$$

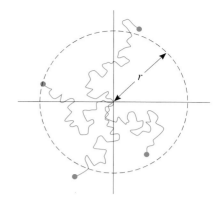

***Figure 2.1*** Random walk of an ion in water.

where $D$ is a constant called the **diffusion constant** which is characteristic of a particular type of ion and which also depends upon temperature. Table 2.1 lists the diffusion constants in water at 25 °C for some of the ions important in living cells. Because the average distance moved by a diffusing ion grows only with the square root of the time, electrical communication by ionic diffusion may be efficient over short distances but becomes prohibitively slow over long distances. For the $K^+$ ion, $D$ is about $1.96 \times 10^{-9}\,\mathrm{m^2\,s^{-1}}$, so that to diffuse distances of 10 µm, 1 mm and 1 m would take it on average 8.5 ms, 85 s and $8.5 \times 10^7\,\mathrm{s} = 2.7$ years respectively (Equation 2.1). So although cells do at times rely on diffusion for the transmission of electrical signals over distances of a few micrometres within a cell, they must use other specialist mechanisms for longer-range electrical communication (as discussed in Chapter 5).

***Table 2.1*** Diffusion constants and enthalpies of selected ions.

| Ion | Radius/nm | Diffusion constant, $D/10^{-9}\,\mathrm{m^2\,s^{-1}}$ | Hydration enthalpy/$10^3\,\mathrm{J\,mol^{-1}}$ |
|-----|-----------|-------------------------------|---------------------------|
| $Li^+$ | 0.06 | 1.03 | −520 |
| $Na^+$ | 0.096 | 1.33 | −405 |
| $K^+$ | 0.133 | 1.96 | −321 |
| $Rb^+$ | 0.148 | 2.07 | −300 |
| $F^-$ | 0.136 | 1.46 | −506 |
| $Cl^-$ | 0.181 | 2.03 | −364 |
| $Ca^{2+}$ | 0.099 | 0.79 | −1 667 |

In water at 25 °C.

In solution, ions, like any other solute, diffuse from regions in which their concentration is high to those where it is low. Imagine a container of solution divided in two by a partition in which the initial concentration of a particular ion is higher on the left side than on the right, while all other properties of the solutions such as temperature and pressure are the same on the two sides of the partition. Figure 2.2a plots the initial ion concentrations $c(x)$ as a function of the distance $x$ along the container. Now imagine carefully withdrawing the barrier so as not to disturb the solutions. The result is a net diffusion of ions from the left to the right and at some later time the concentration of the ion at a position $x$, $c(x)$, will vary smoothly with the distance $x$ along the length of the container, as shown in Figure 2.2b. At any position $x$ along the container, the average rate at which ions flow in the $x$-direction, $J_x$, is found experimentally to be proportional to the concentration gradient at this value of $x$, so that

$$J_x = -D\left(\frac{dc(x)}{dx}\right) \tag{2.2}$$

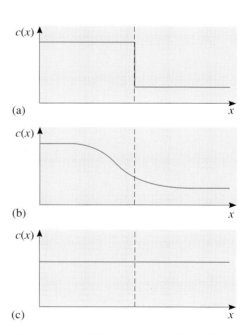

***Figure 2.2*** (a) Ion concentration either side of an impermeable barrier. (b) Diffusion after the barrier has been removed. (c) Final steady state.

This is known as **Fick's** (first) **law** and $D$ is the diffusion constant for this type of ion. Eventually the concentration of the ion becomes the same everywhere in the container, as shown in Figure 2.2c, and there are now no concentration gradients and thus no net diffusive ion flow.

The spontaneous diffusion of ions down an ionic concentration gradient can be explained by a very simple theory in which it is assumed that every ion performs an identical random walk due to collision with the surrounding water molecules, as illustrated in Figure 2.1. Each ion is assumed to act independently and does not interact directly with other ions but only with the surrounding water molecules. This is a plausible assumption because the concentration of water molecules in water corresponds to $55.6 \, mol \, l^{-1}$ so that even at a high physiological ion concentration of $0.1 \, mol \, l^{-1}$ there are still over 500 water molecules for every ion present in the solution.

At every position $x$ in Figure 2.2b each ion is equally likely to move in either direction. However, if we consider a plane perpendicular to the $x$-direction at a particular position $x$ there are more ions immediately to the left of the plane than to the right. Thus if all ions perform identical random walks more ions will cross the plane from left to right than from right to left, simply because there are more ions on the left than on the right. This explains the diffusive flow of ions down the concentration gradient from left to right. Note again that no ion–ion interactions are needed to explain simple diffusion down an ionic concentration gradient.

▷ How would you expect a concentration gradient of ions of one type to affect the diffusive flow of ions of another type?

▶ According to simple diffusion theory all ions act independently of other ions so that the theory predicts that the diffusion of one type of ion proceeds independently of the concentrations of ions of all other types. The total diffusive flow of ions in solution then becomes simply the sum of the diffusive flows of each type of ion independently.

A different type of ion motion is found if, instead of purely random buffeting from surrounding water molecules, an ion is also acted upon by a steady force in a fixed direction. When a free particle is acted upon by a constant force, the particle accelerates at a constant rate so that it travels faster and faster (you may recognize this as a consequence of Newton's second law of motion). However as we have just described, an ion in solution is not free, because of its collisions with the surrounding water molecules. When a constant force is applied to an ion in solution, an equilibrium is rapidly reached such that the rate at which energy is transferred to the ion by the applied force is balanced by the rate at which the ion transfers energy to the surrounding water molecules. Superimposed upon its random walk the ion then drifts in the direction of the applied force $F$ with a constant average velocity $v$. The drift velocity is directly proportional to the force so that

$$v = mF \tag{2.3}$$

where the constant of proportionality $m$ is called the **absolute mobility**. Such motion is called **viscous flow**.

Einstein showed that the absolute mobility $m$ is related to the diffusion constant $D$ by the equation

$$D = mkT \tag{2.4}$$

where $k$ is the Boltzmann constant ($1.38 \times 10^{-23}$ J K$^{-1}$) and $T$ is the absolute temperature. As an example, consider an electric field of strength 100 V m$^{-1}$ applied to the K$^+$ ion in solution at 25 °C. The electric field $E$ is defined as the force acting upon a unit charge, so the force $F$ acting on the ion which has charge $+e$ is $eE$. Substituting this value for $F$ and the value for $m$ from Equation 2.4 into Equation 2.3, we get a value for the drift velocity, $v$:

$$v = (D/kT)eE$$

For the K$^+$ ion, the value of the diffusion constant, $D$, at 25 °C is $1.96 \times 10^{-9}$ m$^2$ s$^{-1}$ (Table 2.1) and at 25 °C, $kT = 4.11 \times 10^{-21}$ J per ion. Therefore the predicted velocity $v$ for the potassium ion is

$$\frac{1.96 \times 10^{-9} \text{ m}^{-2} \text{ s}^{-1} \times 1.6 \times 10^{-19} \text{ C} \times 100 \text{ V m}^{-1}}{4.11 \times 10^{-21} \text{ J}} = 7.62 \times 10^{-6} \text{ m s}^{-1}$$

$$= 7.62 \text{ } \mu\text{m s}^{-1}$$

Thus, whether an ion is freely diffusing or is drifting under the influence of an electric field, its motion is very slow in comparison with the motion of electrons in a metal. This is the greatest difference between ionic currents in living systems and metallic electron currents.

▷ You have already met examples of electron transfer occurring in a living system. Can you recall them?

▶ Photosynthesis and mitochondrial oxidation. However, the mechanisms are totally different from those that occur in metallic conduction.

A second important difference between metallic conduction and ionic current is that while the former is exclusively by electrons, the latter may be carried by many different types of ion. This allows for very subtle effects based on **selectivity** between different ions. A metallic conductor can only be more or less resistive to the electron flow. But in ionic conduction there is the possibility that a conducting path may allow the passage of one type of ion but block the passage of another. The selectivity between different ions may be chemical and caused by the differing binding strengths of the different ions to a given site, or it may be electrical with an attraction or repulsion that depends on the radius of the site and its charge. Consider a site of radius $R_1$ and charge $-q_1$ and a bare ion of radius $R_2$ and charge $+q_2$ (Figure 2.3). The maximum force of attraction between the ion and the site is given by

$$\frac{q_1 q_2}{4\pi\varepsilon(R_1 + R_2)^2}$$

where $\varepsilon$ is the permittivity of the medium (a measure of its polarizability). Therefore for a given ion the force, and thus the selectivity, depends on both the charge and radius of the site. As you will see, this selectivity can lead to highly specific electrical signalling with particular receptors responding only to particular ions.

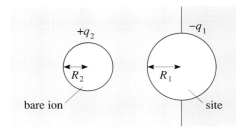

**Figure 2.3** A bare ion adjacent to an oppositely charged site.

▷ A solution of KCl at a concentration of 0.1 mol l$^{-1}$ and at 25 °C is subject to an electric field of 100 V m$^{-1}$. Estimate the current in amperes flowing across an area of 1 mm$^2$ perpendicular to the field due to the K$^+$ ion motion. (One ampere (A) is defined as a flow of charge of 1 coulomb (C) per second.)

▶ To make the calculation, consider a metre cube of solution ($= 10^3$ litres) with one set of its edges parallel to the electric field. If the cube contains $n$ $K^+$ ions each bearing a charge $+e$, and moving at an average speed $v$, then charge passes through the cube faces perpendicular to the field at a rate of $nev$ coulombs $s^{-1}$, so the current is $nev$ A $m^{-2}$. In the present case, $n = 0.1 \times 10^3 \times 6.02 \times 10^{23}$, $e = 1.6 \times 10^{-19}$ coulomb and $v = 7.62$ μm $s^{-1}$, so that the current becomes

$$(6.02 \times 10^{25}) \times (1.6 \times 10^{-19}) \times (7.62 \times 10^{-6})\ C\ s^{-1}$$

$$= 73.4\ A\ m^{-2}$$

The current through 1 $mm^2$ ($= 10^{-6}\ m^2$) is therefore 73.4 μA.

### 2.2.2 Electrostatic self-energy of ions

Before leaving the topic of ionic electricity we need to discuss one further property of ions in solution which is important in understanding the electrical properties of cells. This is the **electrostatic self-energy** which each individual ion possesses. Most electrical energies arise because of the interaction of two or more charges. However, even a single isolated charged particle has an electrostatic self-energy which may be thought of as either the energy required to charge the particle or equivalently as the energy stored in the electric field that surrounds it, as it does every charged particle. An isolated and uncharged conducting particle is at a voltage $V = 0$. If now a charge $Q$ is transferred to the particle, its voltage rises to $V$ where $V = Q/C$, $C$ being the electrical capacitance of the particle. The electrostatic self-energy of such a charge is given by $QV/2$. Changing the environment of an ion will change its self-energy. Table 2.1 lists the **hydration energies** of some common ions – that is, the change in their energy (strictly enthalpy – see Book 2, Chapter 3) when they are transferred from air to water.

The hydration energy is predominantly an effect of a lowering of the self-energy of the ion when it is transferred to water. Water molecules have a large electric dipole moment and the water molecules around an ion are partly polarized by the electric field of the ion to point away from a cation and towards an anion. As an example, Figure 2.4 shows a cation surrounded by water molecules. Although the water molecules are in continuous motion, at any time the electric dipole moments of the water molecules near the cation, which are represented by arrows in Figure 2.4, tend to point away from the cation. The negative voltage created by the predominantly negative ends of the water dipoles near the cation leads to a reduction of the voltage at the surface of the ion and thus a drop in its self-energy. The smaller the ion the greater the electric field at its surface and the greater the local polarization of the water dipoles and hence the greater the hydration energy, as shown in Table 2.1. As you will see later in this chapter the reason why a cell can maintain a difference in ionic concentrations across the cell membrane without a prohibitively large energy cost is a direct consequence of the rise in electrostatic self-energy of small ions on moving from the highly polarizable water to the poorly polarizable lipid membrane.

Examination of Table 2.1 reveals another strange fact that is due to ion hydration. We might expect that $K^+$, which has more than twice the ionic radius of $Li^+$, would diffuse more slowly than $Li^+$. In fact its diffusion constant is about twice as large. This is because of the electrical attraction of the water molecules to the surface of the ion. As discussed above, the smaller the univalent ion, the greater the electric field at its surface and the more powerfully it attracts a shell of water

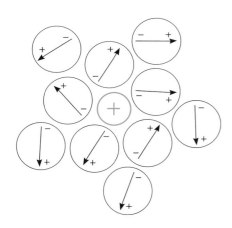

**Figure 2.4** A cation surrounded by water molecules showing orientation of the water dipoles.

molecules about it. Li$^+$ is more effective at holding water molecules at its surface than is K$^+$, and when it diffuses it carries more of its water shell with it. Thus, although the radius of an unhydrated Li$^+$ ion is less than that of the K$^+$ ion, its hydrated size including the captive water molecules is larger, so that in water it has a smaller diffusion constant. An understanding of ion hydration is important when we come to discuss the operation of ion channels which have the role of allowing ions to cross the cell membrane in a controlled manner.

*Summary of Section 2.2*

Because of continuous collisions with neighbouring water molecules, an ion in aqueous solution performs a random walk. In such diffusion the straight-line distance travelled from the starting point at time zero to the end-point at time $t$ grows with the square root of the elapsed time $t$ and its direction is random. Each ion diffuses independently of all other ions but when a concentration gradient of a particular ion exists there is a net flow of the ions down the concentration gradient, as described by Fick's first law (Equation 2.2). If a steady force is applied to an ion, for instance by the application of an electric field, the ion will drift in the direction of the applied force with a constant average velocity proportional to the strength of the force, as described by Equations 2.3 and 2.4. This drift is superimposed upon the random walk. As carriers of electrical signals, ions in solution travel much more slowly than electrons in a metal.

A biological membrane is a very effective barrier to the passage of small ions because of the very high electrostatic self-energy of such an ion when surrounded by material of low electrical polarizability. When a small ion moves in free aqueous solution it carries with it a shell of electrically attracted water molecules.

## 2.3 Transmembrane gradients in a resting cell

Living cells are isolated from their surroundings by their external cell membrane. A voltage $V_m$ exists across the membrane and the concentrations of the various mobile ions inside the cell may be very different from those outside. Table 2.2 shows this for a frog muscle cell. We shall use the convention that all membrane voltages are defined as

$$V_m = V_{in} - V_{out} \tag{2.5}$$

The frog cell is typical of animal cells in having the inside of the cell at a negative voltage relative to the surrounding fluid, giving a $V_m$ of $-100\,mV$.

*Table 2.2*  Ion concentrations in frog muscle fibre and surrounding tissue.

| Ion | Concentration in fibre/mmol l$^{-1}$ | Concentration in plasma/mmol l$^{-1}$ |
|---|---|---|
| K$^+$ | 124 | 2.3 |
| Na$^+$ | 10.4 | 109 |
| Cl$^-$ | 1.5 | 77.5 |
| Ca$^{2+}$ | 4.9 | 2.1 |
| Mg$^{2+}$ | 14.0 | 1.3 |
| HCO$_3{}^-$ | 12.4 | 26.6 |

We shall discuss how these voltage and ionic gradients are related to each other, how they are generated using metabolic energy and why they are essential for a viable cell. However, first we shall investigate the reason why the cell membrane is such an effective ionic insulator that it will allow the cell to maintain these gradients without leaking. A voltage $V_m$ of $-100$ mV across the cell membrane, which is typically only 5 nm thick, corresponds to an electric field of $10^{-1}/(5 \times 10^{-9}) = 2 \times 10^7$ V m$^{-1}$. This is a very large field and would cause the electrical breakdown of most solid and liquid materials. How can this extremely thin lipid membrane resist the passage of ions so effectively when the driving forces are so large? It is not a mechanical effect because the ions are so small (see Table 2.1) that they could easily pass between the highly mobile phospholipid molecules that form the bilayer. In fact the energy barrier to the passage of ions across the membrane is electrical. You saw in Section 2.2.2 that to remove a small ion from water and transfer it to air requires a high energy, essentially because of the low electrical polarizability of the air in comparison with water. Lipid and protein also have low electrical polarizability in comparison with water so that an ion surrounded by lipid molecules would have a high energy for the same reason. To remove a small ion from water and place it in the lipid would cost an energy nearly as large as the hydration energy of the ion. The probability of a small ion surmounting such a large energy barrier to cross the bilayer is negligible so the thin lipid membrane forms a very effective insulator against the passage of small ions.

Such a barrier is essential if the interior of a cell is to maintain a voltage different from its surroundings without the expenditure of large amounts of energy. Within a conducting fluid no voltage differences can exist in equilibrium because, if the voltage at some location is momentarily raised, cations will be repelled from and anions will be attracted to that region by the voltage gradient (electric field) until the voltage difference is exactly neutralized. Transient voltage changes do occur and can be used as signals by living cells before they are neutralized by ion motion, but to maintain a voltage difference in the longer term without great expenditure of energy requires an insulating membrane to prevent ion flow. As you know, within a cell there are of course many insulating membranes surrounding subcompartments such as the nucleus or mitochondria, which means that different parts are maintained at different voltages. Despite these complications of real cells, much can be learnt using a much oversimplified electrical model of a cell as a single body of ionic fluid separated from its surroundings by a cell membrane. Consider, as a particular model of a cell, a 10 μm diameter sphere of univalent salt solution at the typical concentration found in real living cells of 0.1 mol l$^{-1}$, separated from the surrounding fluid by a cell membrane.

An insulating sheet sandwiched between two metal plates forms an electrical capacitor (Figure 2.5). If a voltage difference $V$ is applied between the two plates they will become charged and carry charges of $+Q$ and $-Q$ respectively. The capacitance, $C$, of a capacitor is defined as the ratio of the charge stored to the voltage applied between the plates, i.e. $C = Q/V$ (which is a rearrangement of the equation introduced in Section 2.2), so that

$$Q = CV \tag{2.6}$$

In the same manner, the insulating cell membrane sandwiched between the two conducting ionic fluids inside and outside the cell as shown in Figure 2.6 forms an electrical capacitor. The measured capacitance of a typical cell membrane is about $5 \times 10^{-3}$ farad m$^{-2}$ which is large because the membrane is so thin. (The farad is the unit of capacitance.) Our model spherical cell has a radius $R$ of 5 μm, so its a membrane surface area is

$$4\pi r^2 = 4 \times 3.14 \times (5 \times 10^{-6})^2 \text{ m}^2 = 3.14 \times 10^{-10} \text{ m}^2$$

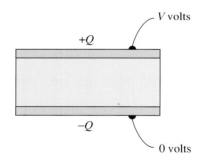

**Figure 2.5** An electrical capacitor.

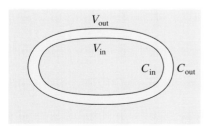

**Figure 2.6** A living cell as an electrical capacitor.

Thus the membrane capacitance is

$$3.14 \times 10^{-10} \times 5 \times 10^{-3} \text{ farad} \approx 1.6 \times 10^{-12} \text{ farad}$$

Thus, to create a voltage of 0.1 V across the membrane requires the transfer of a total charge of $1.6 \times 10^{-12} \times 0.1 = 1.6 \times 10^{-13}$ coulomb across it (Equation 2.6). Since each univalent ion carries a charge of $1.6 \times 10^{-19}$ coulomb, this corresponds to the transfer of about $10^6$ univalent ions. This is a very small number when you think that the model cell contains about $3.15 \times 10^{11}$ cations and the same number of anions. Thus to create a transmembrane voltage difference of 100 mV requires the transfer across the membrane of only 1 in every 300 000 cations contained within the model cell, which is so small as to be undetectable as a concentration change. Thus large transmembrane voltage differences may be created by the transfer of small numbers of ions across the membrane.

### 2.3.1 The Nernst potential

In Section 2.2.1 we showed that ions in solution will spontaneously move down their concentration gradient by diffusion and they will also exhibit viscous flow up or down an applied voltage gradient, depending upon whether they are anions or cations. If only a concentration gradient or a voltage gradient exists we can determine the direction of ion flow using these simple rules. However, in real life there usually exists simultaneously both an ion concentration gradient and a voltage gradient. In order to determine the direction of ion flow in this more complex situation we need to be guided by physical theory. Of particular importance is the case in which voltage and ion concentration differences both occur across the cell membrane. If the membrane is made permeable to one type of ion, by the opening of a pore for example, we need to know whether that type of ion will flow into the cell or out of it. In this section we take the first step by determining the size of the voltage difference across a membrane that will prevent ion flow through a membrane pore due to a given ionic concentration difference across the membrane. This voltage difference is known as the *Nernst potential*. We derive an expression for the Nernst potential below.

Imagine a solution consisting of water as the solvent and a single salt XY as solute which dissociates completely to form $X^{z+}$ and $Y^{z-}$ ions where $z$ is the ionic valency. Within this fluid one region is isolated from the rest by a membrane as shown in Figure 2.6. Assume that the membrane contains pores that will readily permit $X^{z+}$ ions to pass through but will block the passage of $Y^{z-}$ ions. Initially the ionic concentrations are assumed to be the same on both sides of the membrane, but consider what happens when a constant voltage difference $V_m = V_{in} - V_{out}$ is applied across the membrane. A charge of $q$ coulomb located at a position where the voltage is $V$ volts has an electrical energy $U = qV$ joules. The positively charged $X^{z+}$ ions each have a charge $ze$ and so an electrical energy $U_{in} = zeV_{in}$ on the inside of the membrane and $U_{out} = zeV_{out}$ outside. We shall take $V_m$ as negative so that the cations can lower their electrical energy by moving through the pores across the membrane from the outside to the inside. However, such a transfer of ions will lead to an increase in the concentration of $X^{z+}$ ions inside and a decrease in concentration outside. The imbalance in concentration of positive ions on the two sides of the membrane will lead to a diffusive contraflow of these ions from inside to outside.

▷ Can you estimate when the net transfer will cease?

▶ It will stop when the rate of flow of $X^{z+}$ from outside to inside down the voltage gradient is equalled by their flow rate from inside to outside down the concentration gradient.

In the equilibrium steady state in which there is no further ion flow we can write down the probability of finding the cation at each side of the membrane by using the Boltzmann equation, which states that in equilibrium at temperature $T$ the probability $P_i$ of finding a given particle in a state with energy $U_i$ is given by

$$P_i = Ae^{-U_i/kT} \tag{2.7}$$

where $A$ is a constant and $k$ $(1.38 \times 10^{-23}\,\mathrm{J\,K^{-1}})$ is the Boltzmann constant. This is one of the most important relations in understanding the equilibrium properties of ions. Thus, in equilibrium at temperature $T$ the ratio of the probabilities $P_{out}$ and $P_{in}$ of finding a $X^{z+}$ ion outside and inside the membrane is given by

$$\frac{P_{out}}{P_{in}} = \frac{e^{-U_{out}/kT}}{e^{-U_{in}/kT}}$$
$$= e^{(U_{in}-U_{out})/kT}$$
$$= e^{ze(V_{in}-V_{out})/kT}$$

This expression may be written in a more useful form by using the fact that the natural logarithm (that is, the log to the base e) is the inverse of an exponential in the sense that if $X = e^Y$, then $Y = \log_e X$. Taking the natural logarithms of both sides and rearranging, we obtain

$$V_{in} - V_{out} = \frac{kT}{ze}\log_e\left(\frac{P_{out}}{P_{in}}\right)$$

We may now replace the ratio of the two probabilities of finding an ion outside and inside the cell by the ratio of the ionic concentrations outside ($c_{out}$) and inside ($c_{in}$) the cell so that we finally obtain

$$V_{in} - V_{out} = \frac{kT}{ze}\log_e\left(\frac{c_{out}}{c_{in}}\right) \tag{2.8}$$

This relationship connects a voltage difference across a membrane and the difference in concentration of a given ion across the same membrane in the steady state, when only this one ion may penetrate the membrane and there is no net ion flow. The voltage difference across the membrane is known as the **Nernst potential** and written $V_N$. The equation that defines the Nernst potential becomes

$$V_N = \frac{kT}{ze}\log_e\left(\frac{c_{out}}{c_{in}}\right) \tag{2.9}$$

The same expression may be written in terms of a logarithm to the base 10 using the relation $\log_e(X) = 2.303\log_{10}(X)$ so that it becomes

$$V_N = 2.303\frac{kT}{ze}\log_{10}\left(\frac{c_{out}}{c_{in}}\right) \tag{2.10}$$

Note that the concentrations may be expressed in any units, as only a ratio of the concentrations is required, but that the voltages must be expressed in volts. At 25 °C the factor $(kT/ze)$ has a magnitude of 0.0257 V for a univalent ion. The

Nernst potential is the same for all ions of the same valency, independent of their diffusion constant. This is because it refers to the situation of no net flow so that their speed of motion, as measured by their diffusion constant, is not relevant. Transmembrane ion channels that are highly selective for particular ions are common in biology and cells often control the transmembrane voltage by opening and closing such channels. Also, as you will see in the next section, the Nernst potential allows us to express in a simple manner the work that must be expended to move an ion across the membrane. For these two reasons a clear understanding of the Nernst potential is important when discussing the electrical behaviour of cells.

▷ A cell rests in equilibrium with its interior at a voltage 100 mV more negative than that of the surrounding fluid. The membrane of the cell is only permeable to one type of univalent cation. What is the ratio of the internal to external free concentration of that cation?

▶ Equation 2.9 shows that

$$\frac{zeV_N}{kT} = \log_e\left(\frac{c_{out}}{c_{in}}\right)$$

As $V_N = -0.1$ V and $kT/ze = +0.0257$ V,

$$\log_e\left(\frac{c_{out}}{c_{in}}\right) = \frac{-0.1}{0.0257} = -3.891$$

$$\text{thus} \quad \frac{c_{out}}{c_{in}} = 0.0204$$

$$\text{so} \quad \frac{c_{in}}{c_{out}} = 48.96$$

## Summary of Section 2.3

A significant change in the voltage across the membrane of a typical cell is produced by the transfer of a much smaller number of ions than would be required to produce a significant change in the ionic concentration within the cell. If a membrane is permeable to a single type of ion, then equilibrium can exist only if the voltage difference between the inside and the outside of the cell is related to the concentration difference of that type of ion inside and outside the cell as in Equation 2.8. For a given ionic concentration ratio, the voltage difference is known as the Nernst potential as defined in Equation 2.9.

## 2.4 Ion pumps

### 2.4.1 Measuring the work done when ions cross the membrane

No membrane is perfect and all will contain defects which will allow ions to leak across them. Thus any voltage or ionic gradients that are created will decay unless they are actively maintained. To maintain the ionic gradients, cells employ membrane-spanning proteins called **primary pumps**. As an energy source they use the hydrolysis of ATP produced via catabolism. How pumps operate at the molecular level is not yet understood despite the fact that the amino acid sequences of the pump proteins are well known and much is known about their structure and chemistry. The most studied primary pump is the $Na^+$- and $K^+$-dependent hydrolytic enzyme $Na^+/K^+$ ATPase, which will be described in more biochemical detail in Chapter 3. During the hydrolysis of ATP, the $Na^+/K^+$ ATPase enzyme transfers three $Na^+$ ions outward across the membrane and simultaneously transfers two $K^+$ ions inwards across the membrane. Both transfers reinforce the prevailing ionic concentration differences and force ions to move up a concentration gradient when they would otherwise spontaneously flow down such a gradient (as you will recall from Section 2.2.1). To force ion flow up the ionic concentration gradient requires an external energy source and in this section we shall estimate how much work is required to operate an ion pump.

That an external source of work is required to move an ion against a prevailing concentration gradient or against a prevailing electrical energy gradient may be deduced as follows. Equation 2.8 may be rearranged to show that when there is no ion flow and the system is at equilibrium

$$zeV_{in} + kT \log_e (c_{in}) = zeV_{out} + kT \log_e (c_{out}) \tag{2.11}$$

The **electrochemical potential** $g$ of an ion is a quantity defined in Equation 2.12 for an ion of type $i$ at a position where the voltage is $V$ and the concentration of that type of ion is $c_i$. The term $g_o$ is a constant which depends on the solvent and on temperature and pressure.

$$g = g_o + zeV + kT \log_e (c_i) \tag{2.12}$$

(Notice the relationship of this equation for $g$ to Equation 2.11.) Under conditions of constant temperature and pressure (which is the case for most living systems) the electrochemical potential of an ion measures the stored capacity of the ion to do work in its present state. For any change in the state of the ion, such as crossing the membrane, a drop in its electrochemical potential measures the maximum amount of work that can be done *by* the ion during the change, while a rise in the electrochemical potential is a measure of the minimum work that must be done *on* the ion by an external agency for the change to occur. For any change in state of the ion to occur without outside intervention, $g$ must decrease. Equation 2.11, which we derived using the Boltzmann equation (2.7), states that in equilibrium the electrochemical potential for the chosen type of ion is the same at both sides of the membrane. Under the conditions of equilibrium, when the membrane voltage $V_m$ is equal to the Nernst potential $V_N$, no work is done *on* the ion or *by* the ion when it crosses the membrane, as would indeed be expected in equilibrium. Thus, the Nernst potential may be defined as the membrane voltage at which the electrochemical potential of the ion to which the membrane is permeable is the same on both sides of the membrane. At any other membrane voltage, the change in the electrochemical potential of the ion when it moves inward across the membrane, $dg_{in}$, is given by

$$
\begin{aligned}
dg_{in} &= ze (V_{in} - V_{out}) + kT [\log_e (c_{in}) - \log_e (c_{out})] \\
&= zeV_m + kT \log_e (c_{in}/c_{out}) \\
&= ze (V_m - V_N) \tag{2.13}
\end{aligned}
$$

In the derivation of this equation we have used the definition of the Nernst potential $V_N$ in Equation 2.9. Equation 2.13 measures the work done *on* or *by* an ion when it crosses the membrane and is expressed very simply in terms of $V_m$ and $V_N$.

As an example of a calculation of the work required to move an ion across the cell membrane we shall take the frog muscle fibre for which the internal and external ion concentrations are listed in Table 2.2. When the cell is at rest, $V_m = -0.1$ V. Transferring a $K^+$ or $Cl^-$ ion involves very little work, as in each case $V_m$ is close to $V_N$ so that $dg_{in}$ is close to zero. In these cases the change in the electrochemical potential due to the voltage difference across the membrane is nearly cancelled by a change of opposite sign in the electrochemical potential due to the concentration change across the membrane. However, to expel an $Na^+$ ion requires work against *both* the voltage and concentration gradients. For the frog muscle cell at 25 °C, $Na^+$ has a Nernst potential of +0.06 V and the membrane voltage is −0.1 V so that to expel an $Na^+$ ion requires a minimum amount of work $dW_{out}$ to be done *on* the ion by an external source where

$$
\begin{aligned}
dW_{out} &= + dg_{out} \\
&= -dg_{in} \\
&= -ze\,(-0.1 - 0.06) \\
&= -1 \times 1.6 \times 10^{-19} \times (-0.16) \\
&= +2.56 \times 10^{-20} \\
&= 6.2 \times (kT)
\end{aligned}
$$

This is equivalent to the electrical work required to transfer a single univalent cation between two regions which differ in voltage by +0.16 V. The hydrolysis of one molecule of ATP to ADP at pH 7 provides energy equivalent to about $12\,kT$ so that each ATP molecule hydrolysed would only supply enough energy to expel a single $Na^+$ ion from the frog muscle cell. As we have said, the $Na^+/K^+$ ATPase pump drives ions two ways across the membrane, expelling three $Na^+$ ions while pumping inward two $K^+$ ions. By exchanging cations across the membrane, the ionic concentrations inside and out may be adjusted with a minimum direct effect on membrane voltage.

The electrochemical potential $g$ defined in Equation 2.12 is the function which allows us to predict the direction of ion flow when both voltage and concentration gradients exist simultaneously. The term $zeV$ measures the influence of the electrical voltage and the term $kT \log_e(c)$ the influence of the ionic concentration. Spontaneous ion flow in a particular direction is possible only if it causes a decrease in $g$. To bring about an ion flow that would result in a rise in $g$ is only possible if it is coupled to an external energy source. Then the rise in $g$ measures the minimum energy the external source must expend to bring the change about.

To summarize the information that is obtained from measuring changes in the electrochemical potential:

If in any change the value of $g$ *decreases* then the change can occur spontaneously and without outside intervention. Work may be extracted from the change which can be applied to some other part of the system.

If in any change the value of $g$ *increases* then the change will not occur without outside intervention. The outside source must supply work at least as large as the change in $g$ for the change to occur.

▷    The ion current through a transmembrane pore selective for a given cation is measured as a function of the applied membrane voltage $V_m$. What can you say about the sign of the current when $V_m$ is less than or greater than the Nernst potential $V_N$ ?

▶ If $V_m$ is greater than $V_N$ then $dg_{in}$ is positive for a cation (Equation 2.13) and the direction of spontaneous flow of cations will be outward. Thus the measured current will be outward. In a similar manner, if $V_m$ is less than $V_N$ then the cation current will be inward. When the membrane voltage is equal to $V_N$ the current will be zero.

### 2.4.2 Conversion of ion gradients by secondary pumps

The ion and voltage gradients created and maintained by the primary pumps are essential to the viability of the cell. They store energy which is used for many purposes. Perhaps the most important of these is to create secondary ionic gradients using the energy stored in the initial ion gradients. To function effectively a cell needs to establish concentration differences across the membrane of many different ions including protons (i.e there must be a pH gradient). Rather than have a primary pump to create each difference, the cell can exchange one concentration or voltage difference for another by means of **secondary pumps** or **ion exchangers**. These are proteins that span the membrane and use the work done *by* one type of ion when it crosses the membrane to power the transfer of another type of ion across the membrane. The mechanism by which the energy gain of one type of ion moving *down* its electrochemical gradient is used to power the motion of a second type of ion moving *up* its electrochemical gradient is not yet known. However, we are able to say that the coupled transfer of both types of ion will only occur spontaneously if the total electrochemical potential of the system is lowered in the process.

As an example, a cell may use a concentration difference of the $K^+$ ion across the membrane to establish a concentration difference across the membrane of another ion $X^+$ necessary for the cell to function. To investigate the energetics of such an exchange assume the two types of ions have valencies $z_1$ and $z_2$ and thus charges of $z_1e$ and $z_2e$ and let the Nernst potentials of the two types of ion be $V_{N1}$ and $V_{N2}$. If the membrane voltage $V_m$ is such that $z_1(V_m - V_{N1})$ is negative, then $dg_{1(in)}$ is negative so that ions of type 1 would flow inward spontaneously if not prevented from doing so by the membrane. On the other hand, if $z_2(V_m - V_{N2})$ is positive, then for ions of type 2 $dg_{2(in)}$ is positive, so that $dg_{2(out)}$ is negative and they would spontaneously flow out of the cell if not prevented from doing so by the membrane. If a pair of the two types of ion can be transferred into the cell at the same time, then this requires no other source of energy, provided that the total electrochemical potential of the two ions decreases when they make the transfer, i.e.

$$z_1(V_m - V_{N1}) + z_2(V_m - V_{N2}) < 0$$

This process, in which the work made available when an ion of type 1 crosses inward is used to power the simultaneous transfer of an ion of type 2 inward, is called a **coupled transfer**. Clearly, it is only possible because of the transmembrane gradient of ions of type 1 established by a primary pump using metabolic energy. A protein that transports both the driving and the driven ion in the same direction across the membrane is known as a **coport** (or symport) while one that transports the driving and driven ions in opposite directions is known as a **counterport**. Again, little is known about the operation of coports and counterports at the molecular level. When they were first found they were thought to be carrier molecules which bodily moved across the membrane carrying the two ions to deliver them to the other side. However, what is known of the structure of these proteins is not consistent with such bodily transfer across the membrane and the primary structure of what is thought to be the membrane-spanning part of the protein is much like that of an ion channel. We shall compare the properties of carriers and channels in the next section. One possibility is that a coport or counterport consists of two neighbouring ion channels which are coupled together mechanically or electrically, such that the spontaneous passage of one type of ion through a channel selective for that type of ion is coupled to the forced transfer of the other type of ion through the second channel selective for the other type of ion.

▷  A counterport uses the drop in electrochemical potential of one type of univalent cation as it crosses the membrane to drive a second type of univalent cation across the membrane in the opposite direction. How does the total electrochemical potential change of the two ions, which results from the transfers, vary with the membrane voltage ?

▶  At all membrane voltages, the electrical energy required to transfer two univalent cations simultaneously across the membrane in opposite directions is zero. Thus varying the membrane voltage should not alter the change in the total electrochemical potential of the two ions when they cross the membrane.

## Summary of Section 2.4

Ions will move spontaneously from a region of high ionic concentration to a region of lower concentration. They will also move spontaneously from a region in which their electrical energy is higher to one where their electrical energy is lower. When a concentration difference and a voltage difference occur together, an ion will move spontaneously from a region in which its electrochemical potential (defined by Equation 2.12) is higher to a region in which it is lower. The change in the electrochemical potential for a given type of ion when it moves across a cell membrane may be written in terms of the Nernst potential and the voltage difference across the membrane, as in Equation 2.13.

An ion of one type may be forced across the cell membrane in a direction that increases its electrochemical potential, provided that the transfer is tightly coupled to the simultaneous transfer of another type of ion in a direction that lowers its electrochemical potential; the sum of the electrochemical potentials of the two ions must be decreased by the simultaneous transfers. A device that allows such spontaneous coupled transfers is known as a secondary pump or ion exchanger.

## 2.5   Ion transport

### 2.5.1   Ion channels and ion carriers

We have seen how the rise in electrical self-energy is an effective barrier to the spontaneous passage of small ions across a lipid bilayer. However, controlled transfer of ions across the cell membrane is essential to the operation of a living cell. Two very different mechanisms are known to facilitate transmembrane transfer, **ion channels** and **ion carriers**. At the simplest level, an ion channel is a water-filled pore that spans the membrane and allows ions to cross the membrane. An ion carrier or **ionophore** is a molecule that binds to a particular ion and then bodily moves across the membrane carrying the ion with it. Here we shall first contrast some of the important features of ion channels and ion carriers and then discuss in more detail the structure and properties of one particular ion channel.

An ion channel in a living cell consists of a large protein or group of proteins which span the membrane and are exposed on both sides, as illustrated in Figure 2.7. They may be divided into **ligand-gated** or **voltage-gated** channels depending on whether the channel opens in response to the binding of a particular **ligand** (that is, a molecule which binds to the protein constituting the channel) or in response to a change in the local electric field – which is often the result of a change in the local transmembrane voltage. The term 'gated' means simply that the binding of the ligand or the change in voltage opens or closes the channel. Most membrane channels discriminate between cations and anions and some discriminate strongly between different types of ion with the same valency. It is common, even among highly selective channels, for as many as $10^6$ or even $10^7$ ions of a given type to pass through a channel during each second it is open. One powerful method of studying the properties of such channels and their constituent proteins is by cloning the DNA that codes for them. This has been achieved for many of the important ion channels. The cloned DNA can then be introduced into the nucleus of a cell that may not normally produce the channels (by techniques described in more detail in Books 1 and 4). The result of this piece of genetic engineering is that channel proteins are synthesized and become inserted into the cell membranes, where they can be studied in relative isolation, and are found to show properties very similar to their natural counterparts. Although the amino acid sequences of these channel proteins are known, with just one exception the three-dimensional structures of the channels are not. One clue as to their structure can be found in the amino acid sequence, which often contains several stretches of hydrophobic amino acids with a sufficient length to span the membrane in the form of an α-helix. As a result, the currently favoured model of the structure of most ion channel proteins is of an array of parallel α-helices spanning the membrane. The channel is thought to consist of a narrow water-filled pore penetrating through a transmembrane channel protein between the α-helices. The presence of such an aqueous pore would allow a stream of partially hydrated ions to pass across the membrane without too large a rise in their electrostatic self-energy.

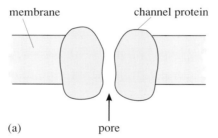
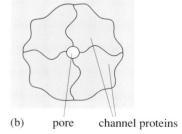

*Figure 2.7* Ion channels, (a) in profile and (b) in cross-section.

An ion carrier or ionophore operates in an entirely different manner. As an example we shall consider valinomycin which is a highly selective carrier of the $K^+$ ion. As shown very schematically in Figure 2.8 it is a small protein which in non-polar solvents is roughly disc-shaped with a height of some 0.4 nm and a diameter of about 0.8 nm. In the interior there is a cavity with six inwardly directed negatively charged ligands that bind the $K^+$ ion. The carrier works by presenting a hydrophobic exterior surface which allows it to dissolve in the non-polar lipid of the membrane. Because of its high self-energy in non-polar solvents, the bare ion will mainly be found in the aqueous phase, but the loaded carrier with its hydrophobic exterior will mainly be in the lipid. In fact, when the loaded carrier has free access to either phase, its concentration in the lipid phase is found to be about $7.5 \times 10^3$ times that in the aqueous phase. The complex can then collect a

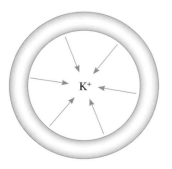

*Figure 2.8* Interaction of valinomycin with a $K^+$ ion.

K$^+$ ion at one membrane face, diffuse across the membrane and deliver the ion to the aqueous phase at the other face. The limiting rate for the transfer is that for the diffusion of the whole complex across the membrane which limits a carrier to a K$^+$ ion transfer rate of about $10^3$ s$^{-1}$.

The feature that most clearly distinguishes the operation of an ion carrier or ionophore from an ion channel is its rate of transfer of ions which at about $10^3$ s$^{-1}$ is much slower than the rate of $10^6$ s$^{-1}$ for the channel. Another marked difference is the temperature dependence. Lowering the temperature can markedly slow the carrier, particularly if the change involves a phase change of the lipid into a more ordered form which inhibits diffusion. Ionophores are found in nature; for instance valinomycin functions in many species as an antibiotic, killing bacteria by causing their outer membrane to leak K$^+$ ions. Carriers of larger molecules will be discussed in the next chapter but, as far as is presently known, ionophores, although important experimental tools, do not play an important part in the normal electrical functioning of living cells. Thus we shall concentrate here on the structure and properties of ion channels.

### 2.5.2   Use of ionic gradients for signalling across the membrane

An important function of the ionic concentration differences established by the primary pumps and secondary pumps is to enable gated membrane ion channels to act as transmembrane signal transmitters and signal amplifiers. Once an electrochemical driving force is established for a particular type of ion, it will spontaneously flow across the membrane whenever an ion channel opens that will allow its transfer. This opens the possibility of transmitting signals, triggered by an external stimulus, across the membrane. In Chapter 5 you will see that it is the ionic concentration differences that provide the power to transmit electrical signals over large distances in excitable cells.

The binding of a ligand to a ligand-gated channel can act as the stimulus that triggers the signal. The binding of a single molecule of messenger ligand to the outer face of a selective transmembrane channel protein can cause the channel to open, resulting in many thousands of ions of a particular type crossing the membrane before the channel again closes. Such a flow may act as an electrical signal by causing a transient change in the local voltage, or as a chemical signal. Such a protein may be considered as a powerful signal amplifier because the arrival of a single messenger molecule can result in the flow of many thousands of ions which themselves may act as messengers. The power supply for the amplifier is the electrochemical gradient established by the membrane pumps. Many parts of the cell may respond to the local entry of ions as they diffuse throughout the cell.

Alternatively, a voltage-gated channel may act as the signal receiver when it opens in response to a change in the electrical field that acts upon it. This change could be due to a local change in the transmembrane voltage or the presence of a charged ligand whose electric field acts on the protein. Complex signalling cascades are clearly possible such as when a ligand-gated Na$^+$-selective channel opens in response to the initiating signal ligand binding to its outer face. The inward rush of positively charged Na$^+$ ions will result in a local reduction in the membrane voltage which could lead to the opening of a nearby K$^+$-selective voltage-gated channel which in turn will lead to the outward flow of K$^+$ ions. Much of the electrical activity of the cell is controlled by the transmembrane ion channels and we shall devote most of the rest of the chapter to their structure and function.

### 2.5.3 Ion channel structure

The three-dimensional molecular structure determined by X-ray diffraction is known for only a few ion channels. The chief reason for this is that it is difficult to form crystals of membrane proteins. A membrane-spanning protein, such as an ion channel, will only maintain its natural shape if its surroundings in a crystal mimic those it experiences when spanning the cell membrane. For example, if exposed over its entire surface to an aqueous fluid it is likely to unfold and take up an entirely different shape to that when spanning the membrane. If the shape of the natural protein is approximated to a circular cylinder it would require a hydrophobic environment over much of its curved surface to mimic the membrane and a hydrophilic environment at its two ends to mimic the intracellular and extracellular aqueous fluids. Much progress has been made recently in this direction and full structures may soon become available.

In the meantime, structural information for some types of channel has been gleaned from electron diffraction experiments performed on closely packed arrays of channel proteins in a two-dimensional lipid sheet. One example is the acetylcholine-activated channel derived from the electric organ of the electric ray *Torpedo*. Looking down on the membrane, the structure the channel has five-fold symmetry about its cylindrical axis. In section perpendicular to the membrane, its outer mouth is wide and extends far into the external medium, its inner cytosolic mouth is equally wide but hardly extends beyond the membrane, and a much narrower pore along its axis extends across the membrane (Figure 2.9).

Now that the complete amino acid sequence of many ion channel proteins has been determined using cloning techniques, it is clear that there are large families of very similar channels. For example, the amino acid sequences of voltage-activated $Na^+$-selective channels cloned from different species are very similar, so it is reasonable to assume that their three-dimensional structures when spanning the membrane will also be similar. Similarly it has become clear that there is a large family of closely related voltage-activated $K^+$ channels.

One of the channels for which the DNA has been cloned is the voltage-gated $K^+$ channel which is particularly important in the transmission of electrical signals. One much studied example is derived from a mutant of the fruit-fly *Drosophila*, known from its characteristic behaviour as *Shaker A* (Figure 2.10). The channel consists of four nearly identical proteins which spontaneously aggregate together across the membrane in a manner like that sketched in Figure 2.7. Each of the protein subunits has a relative molecular mass of about 70 200 and there are 616 amino acids in the chain. Scrutiny of the amino acids present reveals that there are six amino acid sequences composed largely of hydrophobic amino acids, each sequence of a length sufficient to form an α-helix spanning the membrane. Any section of the amino acid chain that is to occupy the hydrophobic lipid bilayer must present an outer surface that is hydrophobic. As you will recall from Book 1, among known protein structures the α-helix and the β-pleated sheet have a secondary structure with suitable hydrophobic exteriors because most of the hydrogen bonds between charged portions of the polypeptide backbone are located internally. Between these lengths there are sequences containing amino acids that carry a charge or an electric dipole which would tend to preferentially occupy an aqueous phase so that the charges and dipoles can be at least partially hydrated.

As a result of these observations, a structure for each of the four channel subunits like that shown in Figure 2.11 is proposed with six membrane-spanning α-helical

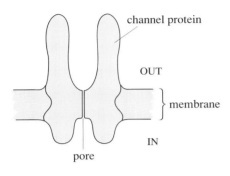

*Figure 2.9* The acetylcholine-activated channel from the electric organ of *Torpedo*.

***Figure 2.10*** The *Shaker* mutation in *Drosophila*. All six flies were anaesthetized with ether; the immobile ones on the left are the normal (wild-type) and those on the right are *Shaker* mutants.

sections connected together by amino acid sequences that are more hydrophilic and situated in the aqueous solutions bathing the faces of the membrane. The sequence between the fifth and sixth helices is highly conserved between different $K^+$ channels and is thought to ring the outer mouth of the pore in the membrane, or even to extend into the pore in a hairpin structure. The evidence for this part of the structure is that very small mutations in the region produce marked changes in selectivity of the channel. The $K^+$ channel is composed of four units such as that shown in Figure 2.11. Further confirmation of such proposed structures is obtained by checking that known binding sites for particular ligands are located on the correct side of the membrane. A very powerful technique is to observe the effects of substitution of the natural amino acid at a particular site along the chain by another amino acid, perhaps with the opposite charge, by appropriately modifying the cloned DNA sequence. We shall discuss the effects of such changes when describing experimental methods for studying channels.

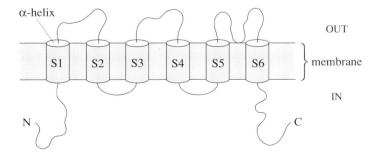

***Figure 2.11*** One protein subunit of the $K^+$ channel from the *Shaker* mutant of *Drosophila*.

It is interesting to compare the amino acid sequence of the $K^+$ channel with that of other voltage-gated channels such as the $Na^+$ channel of the electric organ of *Electrophorus*, the electric eel (Plate 2.1), and the $Ca^{2+}$ channel from rabbit muscle. The *Electrophorus* $Na^+$ channel contains a protein with a relative molecular mass of 208 321 forming a chain of 1 820 amino acids. Along its length there are four regions, each of which has a very similar primary structure to that of the $K^+$ channel protein subunit. All these regions contains hydrophobic sequences that may correspond to the six membrane-spanning $\alpha$-helices of the $K^+$ channel subunit. Thus this part of the $Na^+$ channel is thought to have a marked similarity to the $K^+$ channel except that the four repeats of the same protein that form the $K^+$ channel are linked together in a single protein to form the $Na^+$ channel. The $Ca^{2+}$ channel contains a protein with a structure much like that of the $Na^+$ channel, with a relative molecular mass of 212 018 composed of 1 873 amino acids in the chain. Both the $Na^+$ and the $Ca^{2+}$ channel contain other proteins but the ones described above are thought to form the membrane-penetrating channel. Another similarity is that the fourth $\alpha$-helix, often called S4 (as shown in Figure 2.11), has every third residue occupied by an amino acid that would bear a positive charge in an aqueous environment. The same is true in each repeat of the proposed six helices in the $Na^+$ and $Ca^{2+}$ channel proteins. Much attention has been focused on these sections as the regions upon which the membrane voltage acts in voltage-activated ion channels and, as we shall see, there is some supporting evidence for this.

From such a comparison of the different channel proteins there is evidence of a strong family resemblance between different voltage-activated ion channels and also speculation that the $Na^+$ and $Ca^{2+}$ proteins may have evolved from the simpler $K^+$ structure. The full three-dimensional structure of one channel – the porin channel – has been determined by X-ray diffraction. This channel, which is mentioned in Book 1, and of which you will learn more in the next chapter, belongs to another family and, judging from its conductivity, it has a much larger bore than the channels we have discussed. Much to everyone's surprise, the membrane-spanning part of this channel does not consist of parallel $\alpha$-helices but is composed of the other well-known hydrophobic protein structure, the $\beta$-pleated sheet. There may well be many other surprises such as this when the first full structure of a voltage-gated ion channel is determined.

## Summary of Section 2.5

The transfer of ions across the cell membrane may be facilitated by ion channels or by ion carriers. The two mechanisms are distinguished experimentally by their very different speeds of transfer and their different dependence upon temperature.

A common trigger to initiate biological action by a cell is the opening of a transmembrane ion channel allowing particular types of ions to enter the cell. The channel opening may be caused by a change in the voltage across the membrane or by the binding of a particular signal molecule to the channel protein. Such a trigger will only function if the ionic concentration difference required to drive the transmembrane ion flow is previously established by primary or secondary pumps.

The amino acid sequences of many important transmembrane channel proteins are known and there is much speculation about their structure based upon their sequence. To date, only for very few of these important proteins has the three-dimensional structure been obtained experimentally.

## 2.6   Modelling

Modelling the operation of biological entities such as ion channels is important for at least two reasons. The first is to test our understanding of the underlying mechanisms that lead to its observed properties. It is not sufficient to catalogue the properties of ion channels as revealed by particular experiments. If we are to understand the system in the sense of being able to predict the outcome of situations not yet the subject of experiment, we must enquire whether the experimentally determined properties are consistent with current knowledge of molecular behaviour. In this manner, knowledge of the behaviour of biological molecules may be expanded. The second reason for modelling is to direct future experiments. One fruitful technique is to make a simplified model that incorporates many of the known features of the system, and then to predict further properties based upon the workings of the model which may be tested experimentally. By constant iteration between theoretical models and experimental testing of the models, a more accurate model will evolve.

### 2.6.1   Modelling ion channels

Many well-established properties of ion channels, some of which are listed below, cannot be explained in detail at present. These puzzles may be partially resolved when the structure of the channel protein spanning a membrane has been determined.

(a)  How does the channel protein provide the low-energy path for the passage of small ions across the membrane that ensures ion transfer rates as high as $10^7 \, s^{-1}$?

(b)  How does the channel distinguish between cations and anions with very similar radii such as $K^+$ and $F^-$ (fluoride), allowing the passage of one but not the other?

(c)  How does the channel distinguish between ions of the same charge but very slightly differing radii such as $K^+$ and $Na^+$ in order to allow one but not the other to pass?

(d)  What determines how much current flows through a channel as the voltage across the channel changes?

(e)  What is the mechanism by which the channels open and close in response to the changes in the transmembrane voltage or to ligand binding?

When discussing the resting cell we saw that the barrier to the passage of small ions across the lipid bilayer is the rise in the electrical self-energy of the ion in a non-polarizable lipid environment. A low-energy ion path requires electrically polarizable material surrounding the ion. In a narrow pore, polarizable groups on the channel wall may substitute for some of the surrounding water molecules. If the open pore is to contain water, the protein surface lining the pore must be hydrophilic and participate in hydrogen bonding with the water within the pore. Similarly the outer surface of the channel protein in contact with the surrounding lipid bilayer must be uncharged and hydrophobic to match the non-polar nature of the lipid bilayer. For these reasons, the picture we currently have of the overall structure of an ion channel (Figures 2.7 and 2.9) is of a narrow water-filled pore (or pores) which penetrates through a large membrane-spanning protein with a hydrophobic outer surface where it meets the lipid bilayer and hydrophilic ends where it interfaces with the aqueous phases at either side of the membrane.

## 2.6.2 Modelling selectivity

One form of selectivity is based on the size of the molecules that can pass through the channel. By studying the maximum size of molecule that does pass, it is possible to establish that some channels are much narrower than others. For example, in excitable tissue such studies have established that the narrowest cross-section of the $K^+$ channel is only $0.3\,nm \times 0.3\,nm$, the $Na^+$ channel is wider at $0.5\,nm \times 0.3\,nm$, while the acetylcholine-activated channel found within the synaptic cleft between two communicating nerves (Chapter 3), which is much less selective, has a minimum cross-section as large as $0.65\,nm \times 0.65\,nm$. The concept of a minimum cross-section in the case of charged molecules must be approached with care as the cross-section may be determined electrically as well as mechanically and the effective size of the molecule will depend on the extent to which it carries hydrating water molecules with it, as discussed in Section 2.2.2.

Another form of ionic selectivity is that dependent on the sign of the ionic charge, that is, whether the ion is a cation or anion. This may be modelled by a channel which has regions round the mouth at each end such as those depicted in Figure 2.9, often called vestibules, which are charged. As a much simplified illustration of such a model, let us imagine that each end of a non-selective pore is ringed with negative acidic sites and bathed at both ends by a solution of univalent cations and anions. Such a negatively charged ring will produce a negative voltage within the vestibule which will attract cations into and repel anions from the vestibule. If the voltage in the vestibule is $V_v = -118\,mV$, the energy of a univalent cation in that region will be lowered by $-eV_v$ (since $U = qV$: Section 2.3.1). Equation 2.8 then predicts that the concentration of univalent cations near the channel end will be *increased* over that in the bulk solution, where there is no negative voltage, by a factor of $e^{-eV_v/kT} = 100$ at $25°\,C$. The univalent anion concentration will be *reduced* by the same factor and so the ratio of the number of univalent cations to anions near the pore mouth will be about $10\,000:1$, leading to strong cation-selectivity of the channel even when the pore connecting the two vestibules is non-selective. This may be categorized as selectivity by ion availability. As you will see, there is some evidence of charged rings near the ends of channels.

A much more difficult problem is modelling the selectivity that many biological channels show between different ions bearing the same charge. As you saw from Table 2.1, the bare diameter of the $Na^+$ ion is $0.19\,nm$ while that of the $K^+$ ion is $0.27\,nm$ and yet some $K^+$ channels, such as that at the node of Ranvier in frog nerve, pass more than 100 times the number of $K^+$ ions as $Na^+$ ions for the same concentration gradient of each ion. It is difficult to believe that such selectivity can be based upon a mechanical filter acting on ions that differ in diameter by only $0.08\,nm$. This is particularly so when it is remembered that the channel is subject to thermal vibration and that the ions carry tightly bound water in solution. There is as yet no satisfactory model for this kind of selectivity.

## 2.6.3 Modelling current flow

Modelling the current flow through an ion channel is difficult even using the most basic models. To illustrate this we shall consider the simplest theory which makes the rather drastic assumptions that the channel is entirely passive and does not react mechanically or electrically to the passage of an ion, and that diffusion within a pore is similar to that in bulk solution. We shall further assume that the ions flow through the pore independently and do not interact with other ions within the pore. When there is no voltage difference between the ends of the pore and the pore is selective for a single type of ion, the predicted flow through is simply proportional to the difference in concentration between the ends of the channel as you would expect. However, when there is a voltage between the ends of the pore as in nature, the predicted expression for the flow is complicated. The expression for the predicted inward flow $J_{in}$ of ions per second through a particular transmembrane pore which is selective to a single type of ion becomes

$$J_{in} = A \left( \frac{zeV_m}{kT} \right) \left( \frac{c_{out} - c_{in}e^{zeV_m/kT}}{e^{zeV_m/kT} - 1} \right) \tag{2.14}$$

where $A$ is a constant number characteristic of a particular type of channel. Figure 2.12a plots the prediction of Equation 2.14 for three values of $c_{out}$ with a $c_{in}$ value of $0.1 \, mol \, l^{-1}$. You can see that the flow is not linearly proportional to the membrane voltage except in the special case when $c_{in} = c_{out}$. This at least agrees with experiment in that ion transfer rates for most channels produce non-linear curves when plotted against membrane voltage. As an example the current through an assembly of $K^+$ channels in the giant nerve axon of the squid is shown in Figure 2.12b.

▷ What is significant about the membrane voltages at which the currents represented by the three curves in Figure 2.12 pass through zero?

▶ When the current is zero the membrane voltage is the Nernst potential for the particular internal and external ionic concentrations represented. Using Equation 2.9 with $c_{in} = 0.1 \, mol \, l^{-1}$ and $c_{out} = 0.1, 0.05$ and $0.01 \, mol \, l^{-1}$ the predicted values of $V_N$ are 0, $-17.8$ and $-59.2 \, mV$ respectively.

More recent theoretical investigation has shown that many of the assumptions of the simplest theory are unlikely to hold. Among these are the assumption that the channel is static and does not fluctuate due to thermal excitation or react transiently in its mechanical or electrical structure to the very large electric fields generated by the ion while passing through the channel. Also all the theory used to date assumes that the system studied is at or near equilibrium so that equilibrium or near-equilibrium thermodynamics may be applied. Within a narrow pore, where ions may dwell for as little as $10^{-6} \, s$ while making their transfer it is doubtful if the system is near equilibrium. For example it is doubtful if a meaningful electrochemical potential can be defined for an ion while it is within the pore. Using fast modern computers it may be possible in the future to simulate the conditions inside a pore more accurately. The present situation is that predicted current–voltage relations, such as the simple model we discussed, can act as a guide as to what to expect experimentally, but to use this information to understand the operation of ion channels at the molecular level must await improved theoretical models.

▷ Show that Equation 2.14 predicts that there will be no ion flow across the membrane when the membrane voltage is equal to the Nernst potential for that type of ion.

▶ The predicted ion flow is zero when the term $(c_{out} - c_{in}e^{zeV_m/kT})$ in Equation 2.14 is zero, i.e. when

$c_{out} = c_{in}e^{zeV_m/kT}$

Taking the natural logarithm of both sides and rearranging, this becomes

$$V_m = \frac{kT}{ze} \log_e \left( \frac{c_{out}}{c_{in}} \right)$$

which is our expression (Equation 2.9) for the Nernst potential.

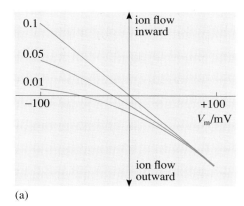

(a)

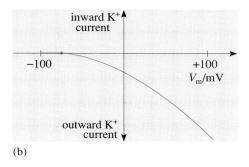

(b)

*Figure 2.12*  (a) Predicted flow of ions through a transmembrane pore at 25 °C. The numbers above each curve are $c_{out}$ values in $mol \, l^{-1}$; $c_{in}$ was kept constant at $0.1 \, mol \, l^{-1}$. (b) Actual measured flow in the squid giant axon.

### 2.6.4  Modelling voltage gating

If a channel is to open and close in response to the membrane voltage, the channel protein must contain charges or electric dipoles that move or change their orientation when the voltage across the membrane changes. Physically this is equivalent to saying that the energy of the protein in the electric field within the membrane is different in the open and closed states. The simplest model is one in which the channel protein has electric charges called **gating charges** which move in the electric field across the membrane.

Consider a particularly simple model with a single charge $+ q$ which can only move abruptly between two stable positions within the channel protein. Assume that the inner position 1 corresponds to a shut channel and the outer position 2 to an open channel. Let the membrane voltage be $V_m = V_{in} - V_{out}$. Let the drop in the membrane voltage experienced by the charge when moving from position 1 to position 2 be a fraction $f$ of the total membrane voltage. Then the electrical energy change when the charge $q$ moves between positions 1 and 2 is $U_2 - U_1$ given by the expression

$$U_2 - U_1 = -qfV_m$$

In the absence of thermal disturbance one would always expect to find the system in its lowest energy state. When $V_m$ is positive the lowest energy state is $U_2$ and when $V_m$ is negative the lowest energy state is $U_1$. Thus in the absence of any thermal disturbance we would expect the channel to exist in its shut state with the charge $q$ in position 1 when $V_m$ is negative and for the channel to switch abruptly to its open state as $V_m$ passed through zero to become positive.

In the presence of thermal disturbance the prediction becomes less exact but we can still compare the relative probabilities of finding the channel in its open and closed states using the Boltzmann relation (Equation 2.7). If the probabilities of the charge $q$ being in positions 1 and 2 are $P_1$ and $P_2$ then in equilibrium at temperature $T$:

$$\frac{P_2}{P_1} = e^{-(U_2 - U_1)/kT} = e^{+qfV_m/kT}$$

However, we know that the charge is always either in position 1 or in position 2, so that

$$P_1 + P_2 = 1$$

Using these two equations we can write the probability $P_2$ of the charge being in position 2 and hence of the channel being open as

$$P_2 = \frac{1}{(1 + e^{(-qfV_m/kT)})} \tag{2.15}$$

Figure 2.13a shows the probability of the channel being open as a function of the membrane voltage $V_m$ as predicted by Equation 2.15 at 25 °C where the product $fq$ is taken to equal $e$, the proton charge (unbroken line). The broken line gives the expected abrupt transition when the thermal disturbance is small. This may be obtained from Equation 2.15 when $T$ becomes small. From Figure 2.13a it can be seen that this simple model can represent the opening and closing of a voltage-sensitive ion channel in response to changes in the membrane voltage.

The changing of the charge structure of real voltage-activated ion channels has been measured when they change from their closed to their open states. These

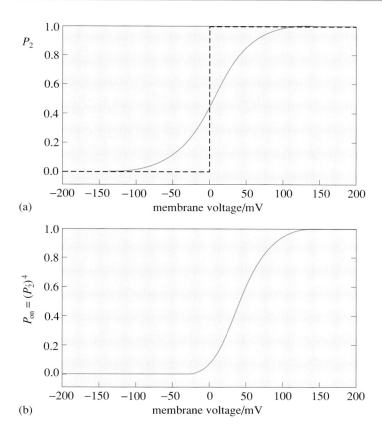

(a)

(b)

*Figure 2.13* (a) Plot of theoretical probability of channel opening against membrane voltage. (b) Shift in the plot as a result of changing assumptions.

measurements of the **gating charge transfer** are important in that they can give us information about the changes in the electrical structure of a channel protein that occur before a channel opens. They should also lead to better understanding of the particular changes in charge structure that must occur when the channel opens. To a first approximation, the measured gating charge transfer for many voltage-activated channels has the dependence on the membrane voltage predicted by Equation 2.15 with particular values for $f$. In our simple model, the characteristic measuring the gating charge transfer is centred on a membrane voltage of zero volts. In real channels such characteristics may be centred at any membrane voltage. Such a shift of the characteristic along the voltage axis could easily be explained by the presence of fixed charged groups in the channel protein which add or subtract a constant voltage from the applied membrane voltage at the location of the gating charge. In our model for a single channel the switch occurs abruptly at a membrane voltage close to zero as the membrane voltage is increased from negative to positive. If the experiment is repeated many times the average probability that the gating charge has transferred between position 1 and position 2 at a particular membrane voltage is a smooth curve such as that in Figure 2.13a.

Our simple model with a single gating charge for the complete channel protein predicts that the probability of gating charge transfer is the same as the probability of channel opening and that both resemble Figure 2.13a. However, the experimentally measured probability of the channel opening is markedly different in that the curve is steeper and shifted toward more positive membrane voltages when compared with the probability of gating charge transfer. Nevertheless, it is possible to construct a theory that is only slightly more complicated than this and which makes predictions much closer to experimental findings.

Consider a channel consisting of a number $m$ of identical subunits each of which has an identical gating mechanism. Let us assume that in any one of the subunits the probability that the gating charge has made a transition from position 1 to position 2 at a membrane voltage $V_m$ is $P_2$, exactly as in the simple theory. The graph of the probability of gating charge transfer against membrane voltage for the complete channel of $m$ subunits then remains the same as Figure 2.13a as each gating charge moves independently. However, if we assume that all $m$ gating charges must make the transition before the channel will open, the probability that the channel is open, $P_{on}$, becomes $(P_2)^m$ which we may write as

$$P_{on} = (P_2)^m = \left( \frac{1}{(1 + e^{(-qfV_m/kT)})} \right)^m \tag{2.16}$$

where we have substituted the value of $P_2$ given by Equation 2.15.

Because the $Na^+$ and the $K^+$ voltage-activated channels in nerve both have four very similar molecular subunits (Section 2.5.3), one might guess that the value of $m$ for these channels would be 4. Figure 2.13b shows the plot of $P_{on}$ from Equation 2.16 as a function of membrane voltage with $m = 4$ and $fq = e$, as in Figure 2.13a. It can be seen that the predicted probability of a channel with four identical subunits being open as shown in Figure 2.13b is a much steeper function of membrane voltage than the probability of gating charge transfer as shown in Figure 2.13a and that it is shifted along the voltage axis. This increase in steepness and the voltage shift between the two curves measuring gating charge transfer and probability of channel opening is similar to that observed in experiments on real channels. However, although it is true that the measured probability that these channels are open as a function of membrane voltage may be obtained very approximately by raising the measured probability of gating charge transfer to some power $m \approx 4$, the agreement between these two functions is far from exact. There is as yet no accepted theory which accurately links gating charge transfer and probability of channel opening as a function of membrane voltage.

### Summary of Section 2.6

The ability to model function is an important test of our understanding of any biological system. Alternating between model building and experimental testing and subsequent refinement of the models is a rapid method of arriving at a sound understanding of the underlying mechanisms involved. Some forms of channel selectivity such as a preference for cations over anions are easily modelled by locating static negatively charged groups near the ends of the channel. However, other types, such as the strong selection of one type of cation over another with a very similar radius, are not yet understood. Current flow through narrow channels is difficult to model kinetically because equilibrium conditions may not exist within the pore. Some static features such as conditions for no flow may be predicted with greater reliability. Measurements of gating charge transfer reflect changes in the electrical structure of the channel between its open and closed states. A simple theory which compares the way in which gating charge and channel conductance vary with the membrane voltage suggest that three or four similar gating charge transfers occur prior to the channel opening.

## 2.7  Experimental investigation of ion channels

Here we shall discuss two different types of experiment that are commonly performed on biological ion channels in order to understand more fully their method of operation. In the first type of experiment the current that flows through the channel is measured as a function of voltage and ion concentrations. This has the advantage of directly reflecting the primary role of ion channels in controlling the flow of ions across the membrane. There is also the hope that even if the operation of the channel is not understood at a molecular level, a detailed knowledge of how it performs can be used within a larger model of cell function. However, the difficulty of constructing realistic models of the flow through ion channels that we discussed above rules out the use of such information as an accurate guide to channel structure. The second type of experiment is aimed directly at an understanding of channel function at the molecular level. This consists of observing the effects on the operation of the channel of the substitution of a particular natural amino acid by another amino acid. These experiments have only recently been possible by altering the sequence of the DNA coding for a cloned channel protein.

### 2.7.1   Measurement of current–voltage characteristics

The flow through an ion channel may be measured either as the flow of individual types of ions (using for example radioactive tracers) or by measuring the current flow. We have shown above that large electrical changes can result from ionic concentration changes that are so small as to be unmeasurable. A single ion channel passing $10^7$ univalent ions per second conducts a current of 1.6 picoamperes ($1.6 \times 10^{-19} \times 10^7 = 1.6 \times 10^{-12}\,\mathrm{C\,s^{-1}} = 1.6\,\mathrm{pA}$). Such a current is readily measurable today. Thus one picoampare corresponds to an ion flow of $10^7/1.6$ univalent ions per second. As 1 mole $= 6.02 \times 10^{23}$ ions (Avogadro's number), this is

$$\frac{10^7}{1.6 \times 6.02 \times 10^{23}}\ \mathrm{mol\ s^{-1}}$$

which is only about $10^{-17}\,\mathrm{mol\,s^{-1}}$. A change in the ionic concentration of only $10^{-17}\,\mathrm{mol\,s^{-1}}$ is certainly not measurable directly and for this reason most measurements are made on current flows. Early experiments could only be performed using naturally occurring specimens and so were limited to easily accessible cells of a sufficient size that electrodes could be inserted. This is why many early experiments on the $Na^+$ and $K^+$ channels in excitable tissue were made using the squid giant axon, which has a diameter as large as 1 mm, much greater than for most nerve axons. Two recent developments have freed experimentalists from these constraints. The first is that intact ion channels can now be introduced into artificial planar lipid bilayers separating two containers of ionic solutions.

Figure 2.14 shows two ionic solution containers between which is sandwiched a very thin plastic sheet. A small hole is formed in this sheet by a hot needle or by passing a spark and an artificial planar lipid bilayer is formed across the hole. This may be done by forming phospholipid monolayers on the surfaces of the solution

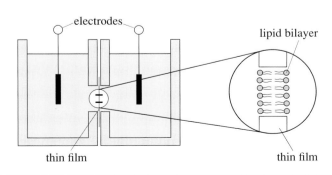

**Figure 2.14**   Ionic solutions separated by an artificial lipid bilayer.

baths when the level is below the small hole connecting the baths. These monolayers are ordered with the hydrophilic head groups in the water and the hydrophobic hydrocarbon tails in the air above the surface. As more solution is introduced into the two chambers and the surfaces of the baths reach the hole a lipid bilayer (see Book 1, Chapter 1) is formed across the hole with the hydrophobic hydrocarbon tails facing each other in the centre plane of the bilayer and the hydrophilic head groups facing outward into the solution baths as in cell membranes. The technique is very flexible and the formation of asymmetric bilayers with different phospholipids in the two halves of the leaflet is possible. Natural ion channels may be introduced into the bilayer by introducing small lipid bilayer vesicles derived from natural membranes into the solution baths. Some vesicles fuse with the artificial planar lipid bilayer and incorporate the channels across the bilayer with their ends exposed to the two solution baths. This set-up is particularly convenient for detailed observation of the conductance of ion channels. The baths may be stirred to ensure a homogeneous ion distribution and the ionic contents of the baths and the voltage applied may be rapidly changed and closely controlled. Because the channel protein may itself respond to the applied voltage, this voltage is often held constant and the current flow across the membrane measured, a technique known as the **voltage clamp**. Pulses of ion current from single ion channels may be observed in this manner and their dependence on the applied voltage and on the chemical composition of the baths studied. This is one of the few occasions when the operation of a single molecule can be readily observed carrying out its biological function.

The second important experimental innovation is the **patch clamp**. Here a small-diameter glass tube is heated, rapidly drawn out and broken off to produce a tapering tube with a diameter of 1 μm or less at its narrow end. The end is flame-polished and the tube is filled with electrolyte solution to form a conducting electrode. The end of the electrode is then gently pressed against the membrane of a cell and slight suction applied as shown in Figure 2.15a. The glass electrode can then form such a high resistance seal to the membrane that only the current flowing into the patch of membrane isolated by the glass tube is measured by the electrode. If, as in Figure 2.15b, a second electrode penetrates into the cell, the ion flow through channels across the small patch of membrane isolated by the first electrode may be studied while the voltage across the two electrodes is varied. In another variant, the suction in the first electrode is increased until the isolated patch of membrane ruptures so that the electrode is open to the whole cell interior. With a second electrode in the fluid external to the cell, as in Figure 2.15c, the combined current for all active ion channels in the cell may be measured between the two electrodes. In a third variation, the electrode is introduced to the membrane and slight suction applied to seal it to the membrane surface. At this point, as shown in Figure 2.15d, the whole electrode is moved backward, which can result in the patch of membrane held by the electrode being torn from the cell. The advantage of this technique (the *inside-out patch*) is that the ionic concentration at the usually inaccessible inner end of the channel may be readily varied as the current between the patch electrode and another electrode in the bathing fluid is measured. Using these techniques, it is now possible to determine the current flow through ion channels that were previously undetectable.

One experiment frequently performed is to measure the current or ion transfer through a channel and determine the transmembrane voltage at which the current is zero. This is the point at which the current reverses its sign and is therefore called the **reversal potential**. For a channel permeable to only one type of ion the reversal potential is of course just the Nernst potential and is a measure of the ratio

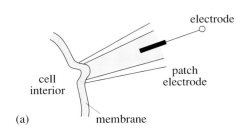

(a)

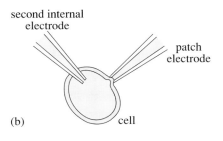

(b)

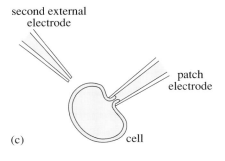

(c)

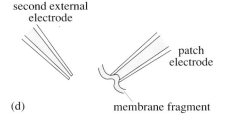

(d)

**Figure 2.15** (a) Making a patch clamp. (b) Inserting a second electrode. (c) Breaking the membrane. (d) The isolated patch ready for inside-out study. (The electrical connection of the electrode is shown schematically in (a).)

of external to internal free concentrations of that ion. Finding a channel for which the reversal potential is the same as the Nernst potential for a particular ion over a wide concentration range of that ion is good evidence that the channel is selective for the ion. The Nernst potential is measured at equilibrium and requires few assumptions in its derivation, so that the interpretation of such experiments is relatively simple. A extension of the derivation for the Nernst potential in Equation 2.9 leads to an expression for the reversal potential $V_r$ of a channel that is permeable to two types of ion, 1 and 2, of the same valency, so that

$$V_r = \frac{kT}{ez} \log_e \left( \frac{p_1 c_{1(\text{out})} + p_2 c_{2(\text{out})}}{p_1 c_{1(\text{in})} + p_2 c_{2(\text{in})}} \right) \tag{2.17}$$

where $p_1$ and $p_2$ are the permeabilities of the channel to the two types of ion. The permeabilities are defined in terms of the net ion transfer rates $\text{tr}_1$ and $\text{tr}_2$ without an applied voltage, such that

$$\text{tr}_1 = p_1(c_{1(\text{out})} - c_{1(\text{in})})$$

and

$$\text{tr}_2 = p_2(c_{2(\text{out})} - c_{2(\text{in})})$$

Measurement of this reversal potential yields the ratio of the permeabilities for the two types of ion $p_1/p_2$ for that channel if the ion concentrations are known. The interpretation of such reversal potentials is not as straightforward as that for the Nernst potential because with two types of ions flowing the system need not be near equilibrium and the theory is on less sure ground. Measurement of the current–voltage characteristic of an ion channel is useful in studying its behaviour under conditions that mimic the real cell but, as we discussed above, is difficult to interpret theoretically in term of channel structure.

### 2.7.2   Amino acid substitution of cloned channel proteins

Current structural models of channel proteins may be tested by the substitution of particular amino acids that appear to play a critical role in the operation of that model. A modified version of the cloned DNA that codes for the modified channel protein is introduced into the nucleus of an oocyte (egg-producing cell). Channels with the modified amino acid sequence are subsequently found in the cell membrane and may then be studied using patch-clamp techniques as described in Section 2.7.1. We shall discuss some experiments performed on the $K^+$ channel from the *Shaker* mutant of *Drosophila*, one subunit of which is shown in Figure 2.11. The first experiments to be described focus on the S4 helix  in which every third residue is arginine or lysine, which are normally positively charged at neutral pH in aqueous solution. Within the S4 helix there are seven of these basic residues. This structure is highly conserved between different $K^+$ channels and because of its high positive charge is thought to be connected with the gating charge which is transferred when the channel opens, as we discussed in Section 2.6.4. The voltage sensitivity of the channel switch-on may be measured by the slope of the graph of the $K^+$ conductance plotted against the membrane voltage. Using cloning techniques, the seven basic residues were substituted one at a time by neutral glutamine residues. Substitution of basic residues 5 and 6 prevented any channel protein being expressed. Substitution of the other basic residues did result in viable channels and their conductance was measured as a function of membrane voltage. Substitution of the third basic residue produced a marked reduction in the slope of the graph of conductance against membrane voltage and also a shift of the entire curve of about 50 mV toward more positive voltages. Substitution of the other basic residues produced only small effects on the slope and small voltage shifts.

Prior to these experiments there was speculation that perhaps the S4 helix was the gating charge and that a bodily movement of this helix partially across the membrane was the structural change that led to channel opening. The large effect of the substitution of the third basic residue does possibly associate the S4 helix with channel gating, but the differences between the effects of substitution of the different basic residues within S4 rules out too simplistic a view of the gating mechanism. Further evidence against a simple association of the S4 helix with the gating charge is that some other $K^+$ channels that have a close similarity in their amino acid sequence to the *Shaker* channel, including the S4 region, are not voltage-gated but ligand-gated. An example is the cyclic GMP-gated $K^+$ channel derived from retinal neurons. (The role of cyclic nucleotides in intracellular signalling is dealt with in the next chapter.) One explanation of the presence of the S4 sequence in a ligand-gated channel is that it could have evolved from a voltage-activated channel.

As a second example, we shall discuss substitution in the structure shown between the fifth and sixth helix in Figure 2.11. This stretch of the protein, called the P region, is thought to line the outer mouth of the channel or even to penetrate into the pore. Evidence for this is that toxins such as dendrotoxin bind to the P region and block the pore when bound and also that substitutions in this region change the selectivity of the $K^+$ channel. An example of how a small number of changes in the amino acid sequence that forms the channel protein can produce large changes in channel function is shown by the removal of only two adjacent amino acids, tyrosine and glycine, from the P region of the *Shaker* $K^+$ channel. The mutated channel shows little selectivity between different cations, in marked contrast to the strong selectivity for $K^+$ ions shown by the unmutated (wild-type) channel. The transfer rate of $K^+$ ions through the two channels remains the same and the mutant channel retains a strong preference for cations over anions. This experiment demonstrates that the strong selectivity shown by many biological channels can depend on very subtle changes in structure.

## Objectives for Chapter 2

After completing this chapter, you should be able to:

2.1    Define and use, or recognize definitions and applications of, each of the terms printed in **bold** in the text.

2.2    Describe in words the chief characteristics of free ionic diffusion as described by Equations 2.1 and 2.2 and of field-driven viscous ion flow as described by Equations 2.3 and 2.4 and utilize these equations to solve simple numerical examples.

2.3    Describe the origin of the electrostatic self-energy of an ion and show how this affects its ability to pass across a lipid bilayer.

2.4    Derive an expression for the Nernst potential (Equation 2.9) and demonstrate its use in defining a connection between the concentrations of a given ion at the two sides of the membrane and the membrane voltage that will result in the steady state if transmembrane pores selective for that ion should open.

2.5    Be able to distinguish between the measurable properties of ion channels and ion carriers or ionophores.

2.6    Distinguish the role of primary pumps in establishing ionic concentration differences across the cell membrane and of secondary pumps in using one ion concentration difference to create a concentration difference of another type of ion.

2.7    Describe in words the significance of the electrochemical potential (Equation 2.12) and be able to predict if the coupled transfer across the cell membrane of a number of ions of different types can happen spontaneously or requires work to be done by an external energy source.

2.8    Describe how ion channels can detect an electrical or chemical signal outside the cell and then transmit a secondary amplified signal across a membrane.

2.9    Explain the purpose and current results of attempting to model some of the important characteristics of ion channels.

2.10    Know of some experimental techniques used to study the operation of ion channels.

## Questions for Chapter 2

### Question 2.1    (Objective 2.2)

An $Na^+$ ion in aqueous solution at $25\,°C$ is acted upon by an electric field $E$ of $100\,V\,m^{-1}$ for a time $t$. What is the minimum time during which the field must act, if the distance the ion moves under the influence of the field is greater than the average distance the ion diffuses in the same time?

### Question 2.2    (Objective 2.3)

To move an ion of charge $+q$ to a region where the voltage is higher by $V$ volts requires energy $U = qV$. A membrane separates two solutions of NaCl with the same concentrations of $Na^+$ ions but between which there is a voltage difference of $100\,mV$. Compare the energy change when a $Na^+$ ion is exchanged between the two solutions with $kT$ a typical thermal energy at a temperature $T$ of $25\,°C$.

### Question 2.3    (Objective 2.4)

From the concentration differences for the $Na^+$, $K^+$ and $Cl^-$ ions across the cell membrane of frog muscle fibres given in Table 2.2, calculate the Nernst potentials for these ions at $25\,°C$ ($kT/e = 0.0257\,V$). If the outside of the cell is maintained at zero volts, what will the voltage inside the cell become in the steady state if the membrane becomes selectively permeable to each of these ions in turn? Compare the voltages you calculate with the voltage of $-100\,mV$ at which the interior of the cell rests. What can you say about the effect of a leak in the resting membrane to each of the three types of ion?

### Question 2.4    (Objectives 2.4 and 2.7)

A cell at $25\,°C$ has an internal $K^+$ ion concentration of $150\,mmol\,l^{-1}$ and an external concentration of $K^+$ ions of $20\,mmol\,l^{-1}$. The interior of the cell is $60\,mV$ more negative than the fluid surrounding the cell. If an ion channel selective for the $K^+$ ion should open across the cell membrane, will $K^+$ ions have a net flow into or out of the cell ?

## Question 2.5   (Objective 2.7)

A counterport operates with the driving ion having a valency $z_1$ and a Nernst potential of $V_{N1}$ and the driven ion having a valency $z_2$ and a Nernst potential of $V_{N2}$. The membrane voltage is $V_m$. The *inward* transfer of a single ion of type 1 provides the energy for the simultaneous *outward* transfer of *two* ions of type 2. Obtain an expression connecting the two Nernst potentials and the membrane potential that must be obeyed if such a counterport is to function spontaneously without an external source of energy.

## Question 2.6   (Objective 2.9)

Draw graphs of $P_2$ and of $P_{on}$ for an assembly of identical channels as a function of $V_m$ in millivolts using the expressions we derived in Section 2.6.4 assuming that $fq = e$, the proton charge, that $m = 4$ and that the temperature $T$ is such that $kT/e$ = 25 mV. Estimate the proportion of channels that are open at the membrane voltage at which half the gating charge has been transferred. What will be the effect on the graphs if $fq$ is increased in size?

# *Intracellular chemical signalling*

## *3.1 Preamble*

This chapter is concerned with the processes by which the cell responds to information arriving at its external membrane in the form of chemical signalling molecules, which can either cross the cell membrane to interact directly with intracellular proteins or bind to receptors located on or in the membrane. The consequence is a cascade of enzyme-catalysed reactions which can dramatically redirect the course of the cell's activity. This chapter builds on the essentially biophysical description of these cellular mechanisms given in Chapter 2, but there we were mainly concerned with ion fluxes and processes describable in electrical terms; here we are firmly in the terrain which biochemistry claims as its own. The regulation of cellular metabolism is a very finely tuned affair, with a multitude of chemical checks and balances to ensure both homeostasis and homeodynamics. It is necessary both to keep the cell's internal 'housekeeping functions' under control whilst at the same time enabling it to respond rapidly to the exigencies imposed by fluctuations in its environment, whether that environment is external to the organism, as for bacteria, or is the other tissues and extracellular spaces which surround any individual cell in a multicellular organism. It has taken biochemists many years to unravel these processes, and there is no simple way to describe them.

In this chapter you will therefore inevitably find yourself in something of a biochemical jungle, full of molecules with unwieldy names. But do not despair; we are not expecting you to remember the fine detail of each biochemical process described. You need only know the general principles of signalling mechanisms, first and second messengers, and the ways in which regulation can be achieved by enzyme cascades in which the activity of each enzyme can be affected not only by the presence of ions or other classical cofactors but also by covalent modification of the enzyme protein itself. In all these complex processes, two general biochemical principles emerge, which will crop up again and again, not only in this chapter but in later ones too, as well as in Book 4. These principles are: (a) the regulatory role of rapid fluxes in the concentrations of calcium ions, and (b) the control of complex cascades by the relatively simple device of phosphorylation and dephosphorylation of particular amino acid residues located at key sites within the polypeptide chains of the enzymes. Keep a cool nerve as you read on, and use these general principles to guide you through the biochemical jungle.

## *3.2 Transport across the cell membrane: an introduction*

Before we launch into the principles and some of the finer details of intracellular signalling pathways it is necessary to sidestep and first consider the transport properties of the cell membrane. These properties are relevant to the forthcoming sections on intracellular communication since the cell membrane is, in many cases, a key player in the transfer of information into cells.

As you will recall from Book 1, phospholipid bilayers are the major component of all biological membranes. They act as a partition between the cell and its surrounding environment and also segregate intracellular compartments such as the nucleus, the mitochondrion and the endoplasmic reticulum. In the previous chapter you learnt about the effective electrical insulating properties of the cell membrane and how ions employ channel proteins in order to traverse this permeability barrier. Obviously, all chemical traffic between the inside and outside of the cell also has to pass through the cell membrane and this passage occurs in different ways; transport is either passive, by diffusion, or actively mediated.

Three factors limit the rate of diffusion of substances across a membrane:

o   the thickness of the bilayer ($l$)

o   the concentration gradient across it ($c_2 - c_1$)

o   the diffusion coefficient ($D_m$).

These are related as follows:

$$J = -\frac{D_m \, (c_2 - c_1)}{l} \tag{3.1}$$

where $J$ is the rate of transport in moles per cm$^2$ per second.

$D_m$ is not the same as the diffusion constant ($D$) that an ion would have in aqueous solution (discussed in Chapter 2, Section 2.2.1), as it depends not only on the size and shape of the molecule, but also on the viscosity of the membrane lipids and on the molecule's solubility in the phospholipid bilayer (which approximates to its solubility in oil). Since the thickness of membranes is not usually known with exactness, it is often more convenient to describe diffusion rates in terms of a permeability coefficient, $p$, as shown in Figure 3.1. Values of $p$ can be determined experimentally from measured rates of transport ($J$) for known concentration gradients, since

$$J = p \, (c_2 - c_1) \tag{3.2}$$

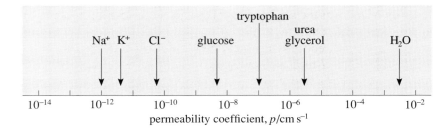

**Figure 3.1**   Permeability coefficients of some ions and molecules in lipid bilayer membranes.

▷   From Figure 3.1, identify substances that cross a lipid bilayer most readily. What chemical characteristics enable them to permeate a cell?

▶   The hydrophobic core of a membrane renders it highly permeable to small non-polar molecules such as urea. Larger hydrophobic substances such as fatty acids and steroids (not shown in Figure 3.1) can also cross the bilayer with ease. Although oil and water are clearly immiscible, the cell membrane has a surprisingly high permeability for water. Its ability to readily penetrate the bilayer is probably a consequence of its small molecular size.

Recently, another mechanism has been proposed which accounts for very high rates of water movement across the membrane of some cells. Membrane-spanning proteins called *aquaporins* appear to form water-selective channels across the phospholipid bilayer and thus allow the rapid diffusion of water between the cell cytosol and the surrounding environment.

▷   Which of the substances in Figure 3.1 have most difficulty in penetrating the membrane? Why?

▶ The permeability coefficients of ions are very low (in the range $10^{-12}$–$10^{-10}$ cm s$^{-1}$) and values for hydrophilic molecules such as glucose are only a couple of orders of magnitude higher (Figure 3.1). This is because of unfavourable interactions with the hydrophobic core of the phospholipid bilayer.

Hence, as in the case of intracellular electrical signalling, the simple diffusion of hydrophilic solutes across the cell membrane between the cytosol and the extracellular milieu (or indeed between various intracellular compartments) is not sufficient to meet the demands of the organism, so that their entry and exit from the cell also require specialist membrane proteins. The structure and functions of these proteins is discussed further in Sections 3.4 and 3.5, but first we shall consider some of the important thermodynamic aspects of membrane transporters.

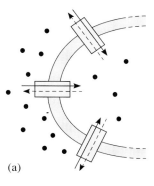

(a)

## 3.3 Transport thermodynamics

### 3.3.1 Passive transport

In the absence of an input of energy, an uncharged substance will traverse the cell membrane until its concentration is the same on both sides of the membrane (Figure 3.2a).

▷ If a charged substance is considered, what other factor affects its movement (assuming there is no input of energy)?

▶ The voltage difference across the cell membrane (discussed in Chapter 2, Section 2.3).

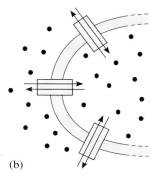

(b)

Equilibrium may be achieved either via simple diffusion through the bilayer or by specific proteins such as the aquaporins. At equilibrium there is no net movement across the bilayer, although molecules still randomly traverse the membrane at equal rates in both directions (irrespective of whether this process occurs by diffusion or by a carrier-mediated mechanism), i.e. there is a continual dynamic exchange of substances between cell compartments and the external medium even after equilibrium has been reached (Figure 3.2b).

Passive processes rely on the presence of concentration gradients across membranes (and also electrical voltage gradients in the case of charged substances such as ions), and these prerequisites impose significant limitations on the magnitude and direction of the exchange of solutes. It is possible to enhance the rate of passive transport by the presence of binding proteins on one side of a membrane. These sequester the substance after it has been transported, thereby maintaining the diffusion gradient, yet permitting the accumulation of a solute in either the intracellular or extracellular space (Figure 3.2c). Examples of this process include: the presence of haemoglobin in erythrocytes (red blood cells), enabling these cells to carry oxygen at high concentrations for delivery to tissues; binding proteins in the periplasmic space (between the cell wall and the cell membrane) of bacteria which concentrate sugars and amino acids prior to their uptake; and fatty acid binding proteins which are present in a number of tissues, promoting the uptake and subsequent utilization of fatty acids during starvation.

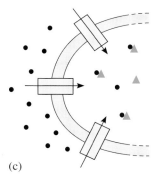

(c)

☐ specific transporter located in cell membrane

▲ binding protein

• solute

***Figure 3.2*** (a) Passive, carrier-mediated diffusion of a solute down a concentration gradient. (b) Dynamic equilibrium: concentrations of solute on both sides of the cell membrane are equal, but there is still exchange between the two compartments. (c) A concentration gradient of a solute across a membrane may be maintained by the presence of binding proteins.

### 3.3.2 Active transport

Not all substances that the cell might require can enter by diffusion down an electrochemical or concentration gradient. For instance, some are at a higher concentration inside than outside. Active transport systems maintain such a concentration difference by promoting the movement of substances against their gradient, thus enabling them to enter (or in other cases to leave) the cell. Such movement is thermodynamically unfavourable so a free energy source is coupled to the redistribution of solutes, causing their accumulation or exclusion from the cell. You have already encountered such active transport systems in Chapter 2, where their role in the distribution of $Na^+$ and $K^+$ ions was mentioned. Indeed, one major function of active transport is to control the ionic composition of the cell (Table 3.1 – an expansion of Table 1.3 in Book 1). The transmembrane voltage and ionic concentration gradients which result from the asymmetric distribution of intracellular and extracellular ions represent an important energy store which can be linked to the transport of other hydrophilic substances. Active transport systems also regulate pH and intracellular volume and are important for generating electrical excitability in nervous and muscle tissue.

**Table 3.1**  A comparison of ion concentrations inside and outside a typical mammalian cell.

| Ionic component | Intracellular concentration /mmol $l^{-1}$ | Extracellular concentration/mmol $l^{-1}$ |
|---|---|---|
| *Cations* | | |
| $Na^+$ | 5–15 | 145 |
| $K^+$ | 140 | 5 |
| $H^+$ | $4 \times 10^{-5}$ (= pH 7.4) | $4 \times 10^{-5}$ (= pH 7.4) |
| $Mg^{2+}$ | 30 | 1–2 |
| $Ca^{2+}$ | 1–2† | 1–2 |
| *Anions** | | |
| $Cl^-$ | 4 | 110 |

* The number of anions should balance the number of cations in both compartments. The deficit, which is particularly large inside the cell, reflects the fact that many other anionic components not included in the table (e.g. $HCO_3^-$, phosphate, nucleic acids, carboxylic acids, proteins) also contribute to the negative charge balance.

† The majority of the intracellullar calcium is present in organelle stores. The cytosolic concentration is only about $10^{-7}$ mol $l^{-1}$ (0.1 μmol $l^{-1}$).

You will recall from Chapter 2 (Section 2.4.1) that ATP hydrolysis provides the energy to drive the asymmetric distribution of ions across the cell membrane. In specific cases, ATP also drives the uptake of substances into intracellular compartments. These are referred to as primary active transport systems, so-called because of their direct dependence on ATP. Secondary active transport is driven by the ionic gradients set up by primary systems. For example, ATP fuels the accumulation of $Na^+$ in the extracellular space and the inward movement of $Na^+$ down its concentration gradient can be coupled to the uptake of glucose, which is cotransported into gut and kidney epithelial (lining) cells.

**Group translocation** is another active transport process which is frequently employed by prokaryotes. Sugars are taken up by coupling their entry to phosphorylation; for example, glucose traverses the cell membrane by a process called **facilitated diffusion** and then is immediately converted to glucose 6-phosphate by the donation of a phosphate group from the glycolytic intermediate phosphoenolpyruvate (see Book 2, Chapter 4) as shown in Figure 3.3.

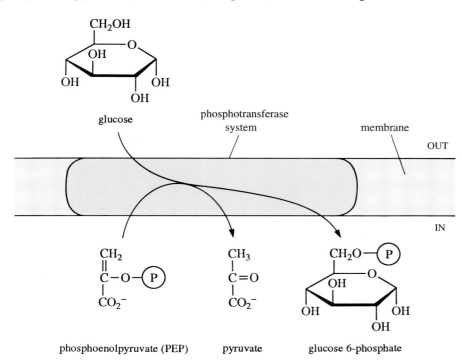

*Figure 3.3* Active transport of sugars into bacteria by group translocation. ('Out' denotes the periplasmic space, between the cell wall and the external membrane.)

Since it is ionized and chemically modified, the phosphorylated sugar is now unable to leave the cell, either by diffusion or via the specific transport system through which it entered, thus maintaining the concentration gradient of the non-phosphorylated sugar from inside to outside. This strategy is akin to that employed by the binding proteins; the trapping of a transported substance, either by a binding association or chemical modification, sustains its rate of transport across a bilayer, as it prevents an equilibrium being reached within the cytosol.

Energy for active transport can also be provided by light. In the archaebacterium *Halobacterium halobium* the purple membrane protein, bacteriorhodopsin, contains a light-absorbing group. The absorption of a photon induces a conformational change in the molecule and this generates an electrochemical gradient by translocating protons into the periplasmic space. This proton gradient is subsequently used to drive the synthesis of ATP. You will recall from Book 2, Chapters 4 and 6, that proton-pumping assemblies are also found associated with the electron transport chains of the inner mitochondrial membrane and chloroplast thylakoid membranes. ATP-synthesizing proton gradients are established in these organelles by the flow of electrons generated by the oxidation of fuels or by the trapping of light energy via chlorophyll excitation and subsequent electron transfer.

## 3.4 Structure and function of transport proteins

How does the phospholipid bilayer behave as a selectively permeable barrier? Consider the molecular architecture of a typical membrane. You will recall from Book 1, Chapter 1, that cell membranes have a fluid mosaic organization; the phospholipids are interspersed with a variety of proteins, including those that traverse the entire bilayer (Figure 3.4).

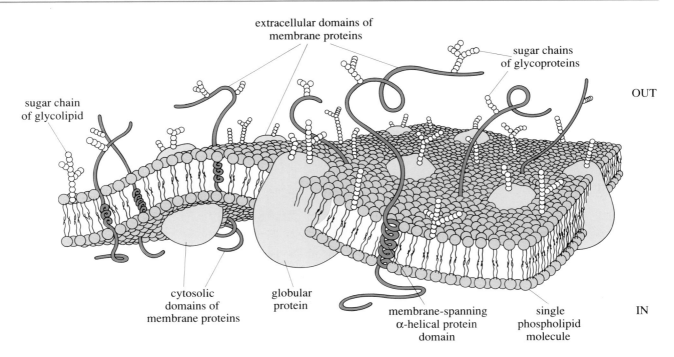

**Figure 3.4** Fluid mosaic model of membrane structure.

Many of these so-called *integral* membrane transport proteins have had their amino acid sequence elucidated and have been subjected to structural analysis. As discussed in the previous chapter, all the membrane polypeptides that have been investigated appear to consist of several hydrophobic domains.

▷ What can be predicted from such a structure?

▶ The hydrophobic domains of transport proteins will readily enter the lipid bilayer to form a transmembrane multi-looped structure. Hydrophilic solutes can therefore traverse the membrane via this continuous protein pathway, avoiding any contact with the bilayer's hydrophobic core.

Recall from Chapter 2, Section 2.5, that there are two distinct modes of ion translocation: carrier proteins and channel (or pore-forming) proteins (Figure 3.5). We can now extend these models to encompass the transport of a diverse range of hydrophilic substances (including sugars and amino acids).

**Figure 3.5** Solute translocation mechanisms. (a) A carrier protein binds solutes with high affinity and stereospecificity. This induces a conformational change which enables solute to be moved from one side of the membrane to another. (b) Channel proteins provide a low-resistance route for solutes (usually ions) by forming an aqueous pore in the cell membrane.

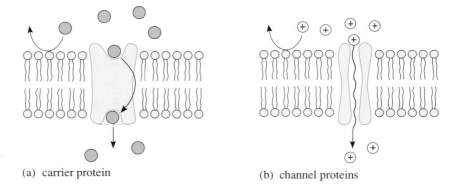

(a) carrier protein

(b) channel proteins

In Chapter 2 you learnt that ion channels play a major role in the electrical functions of living cells. The channels form water-filled pores across the bilayer which may provide a low-resistance route for ions to enter or leave the cell (Figure 3.5b). The direction of ion movement through an aqueous pore is determined solely by its electrochemical gradient (recall the Nernst equation in Chapter 2), so that the coupling of a channel to an energy source would not provide a functional mechanism for driving its movement against the gradient (although it may provide a means of regulation – see Section 3.5.5). You will recall from Chapter 2 that many of the ion channels found in the cell membrane are both highly selective and tightly regulated (gated). They can be opened and closed by a variety of stimuli (mechanical, electrical or chemical) and this is important in transmitting and amplifying information across the cell membrane (discussed in further detail in Sections 3.7–3.9).

However, not all channel-forming proteins are involved in the translocation of ions, and not all undergo such stringent regulation and selectivity. For example, porin (previously mentioned in Book 1), is a protein present in the outer membrane of bacteria (and also in the outer membranes of mitochondria and chloroplasts) which has a large-diameter, unselective pore which is permanently open.

Carrier proteins bind solutes with high affinity and stereospecificity (although there may be a degree of tolerance for the binding of chemically related substances). Binding induces a conformational change in the membrane protein which results in the movement of the solute from one side to the other. The transport process is completed when the solute, having traversed the bilayer, dissociates from the opposing face of the membrane (Figure 3.5a). In the case of active transport, conformational changes may be driven by the hydrolysis of ATP, which can generate electrical and chemical gradients across the bilayer, for example the operation of the $Na^+/K^+$ pump discussed later.

Carriers and channel proteins may be distinguished on the basis of their kinetic properties. The **turnover number**, that is the number of molecules transported per second, is at least 100 times slower for carriers than for channels because the conformational changes involved in carrier-mediated transport impose considerable kinetic limitations. Both types of transporter protein can become saturated (making an easy distinction between protein-mediated transport and simple diffusion), although the saturation of carriers tends to occur at lower concentrations than that of channels (Figure 3.6). These differences in saturation thresholds have been attributed to the differences in the strength of interaction of the solute with the two types of transport proteins; the binding of a solute to a carrier protein is considered to be more tenacious than is the interaction of a solute with a channel protein.

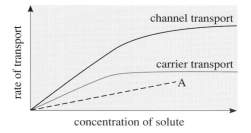

**Figure 3.6** Transport kinetics of channels and carriers. Channels allow faster rates of solute movement which saturate at a higher concentration of solute than do carriers.

▷ What would be the mode of transport of substance A?

▶ A is likely to be transported by simple diffusion since there is no evidence that saturation occurs.

The stereospecific nature of the binding sites on carrier proteins is often compared to that of an enzyme–substrate interaction (which was discussed in Book 2). Indeed, the kinetic parameters of a carrier protein may be quantified in a similar fashion, applying $K_M$ as an index of the affinity of the substrate for the carrier and $v_{max}$ as its maximum activity. It is now recognized that the high degree of selectivity detailed for some of the ion channels discussed in Chapter 2 is associated with the binding of the solute to a region of the protein pore. The formation of such high-affinity, long-lived interactions is incompatible with very high flux rates through the channel since such interactions would restrict ion flow. Perhaps slower channels exist as a result but, because of their decreased rates of transport, they have escaped identification. It is these recent insights into the functioning of transport proteins which have led to a change in the way that biochemists view the operation of carrier proteins and channels; rather than two discrete groups, more complex models are now being considered, and the channel and carrier concepts are now regarded as the extremes of a continuum of transporter mechanisms.

## 3.5 Some important transport systems

### 3.5.1 The sodium/potassium pump (Na+/K+ ATPase)

The **sodium/potassium pump** is present in the cell membrane of virtually all animal cells. It pumps $Na^+$ ions out and $K^+$ ions in to the cell, thereby maintaining concentration differences of about 10–30-fold across the bilayer. It does this at the expense of the hydrolysis of ATP, so it is also an ATPase enzyme.

The purified protein has been identified as a tetramer consisting of two membrane-spanning $\alpha$ chains (which loop backwards and forwards through the membrane) and two smaller $\beta$ subunits which are largely exposed to the outer surface and carry oligosaccharide chains. Chemical cross-links can be readily formed between the two $\alpha$ subunits or between $\alpha$ and $\beta$, but not between $\beta$ and $\beta$, suggesting that the $\alpha$ chains are in contact but the $\beta$s are not (Figure 3.7a). A functional $Na^+/K^+$ pump which, in the presence of ATP, pumps the cations in opposite directions (Figure 3.7b) can be reconstituted within phospholipid vesicles. In the presence of ADP, $P_i$ (inorganic phosphate) and appropriate cationic gradients, the pump can also be driven in reverse and used to make ATP, although this does not occur physiologically. Current estimates suggest that three $Na^+$ ions are pumped out in exchange for two $K^+$ ions entering the cell when one molecule of ATP is hydrolysed, so the ATPase generates an uneven charge distribution (or potential difference) across the cell membrane (i.e. the transporter is **electrogenic**). This appears to be consistent with thermodynamic principles; the energy liberated when ATP is hydrolysed is sufficiently large to provide the free energy required to move the cations against their respective gradients (see Chapter 2, Section 2.4.1).

The spatial organization of the ATPase in the membrane has been investigated using so-called erythrocyte 'ghosts'. Erythrocytes (Figure 3.8a) immersed in a hypotonic salt solution (that is one whose osmotic pressure is lower than that of the inside of the cell) swell and become leaky (Figure 3.8b). The haemoglobin is lost and a pale ghost cell is left with an internal composition in equilibrium with the surrounding medium. When the cells are resuspended in isotonic solution (that is one with the same osmotic pressure as the contents of the cell) the membrane reseals and again functions as a permeability barrier (Figure 3.8c). In this way, the internal composition of the ghost can be varied. Experimental investigations have established that $Na^+$ and ATP must be present inside the cell, whereas $K^+$ must be outside to enable transport to occur. ATP is not hydrolysed unless $Na^+$ and $K^+$ are transported; tight coupling with transport is achieved by the transient phosphorylation (by ATP) of an aspartate residue on the $\alpha$ subunit (Figure 3.7c), which occurs when $Na^+$ is present. This promotes a conformational change of the protein, causing the movement of $Na^+$ ions out of the cell. The phosphorylated aspartate residue is subsequently hydrolysed in the presence of extracellular $K^+$ ions which can now bind to a site exposed by the movement of $Na^+$ ions. $K^+$ ions are then translocated to the interior of the cell as the ATPase returns to its original conformation (Figure 3.7b). (Note that $Mg^{2+}$ – not shown in the figure – is an additional cofactor required by all ATPases.)

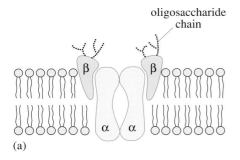

oligosaccharide chain

(a)

***Figure 3.7*** (a) The structural organization of $Na^+/K^+$ ATPase. The sodium/potassium pump is composed of two large membrane-spanning $\alpha$ subunits and two small $\beta$ subunits which carry oligosaccharide chains.

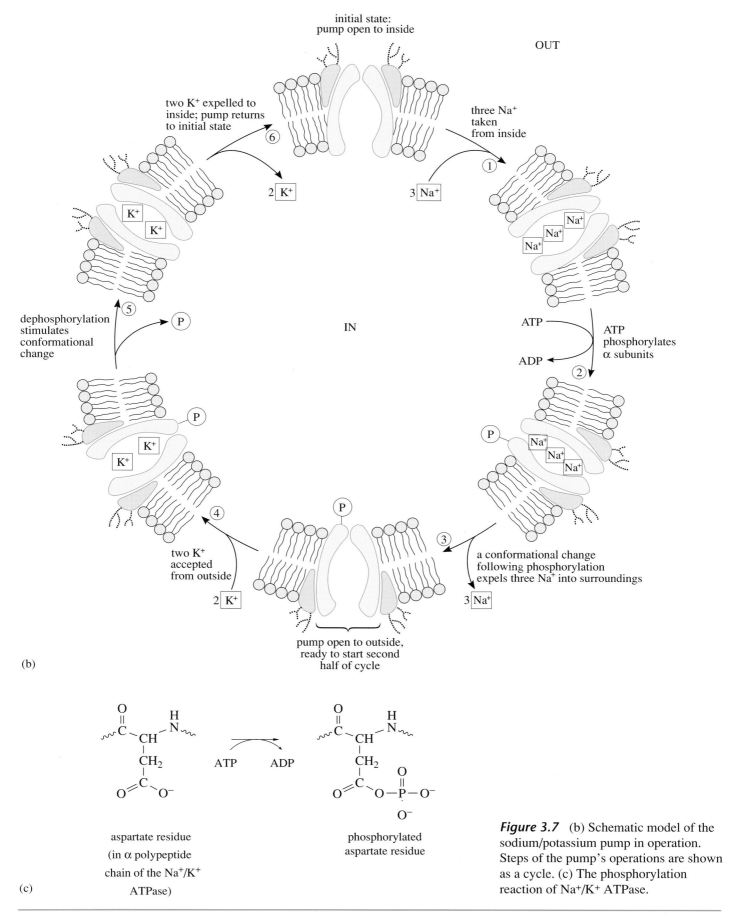

initial state:
pump open to inside

two K⁺ expelled to
inside; pump returns
to initial state

⑥

2 K⁺

three Na⁺
taken
from inside

①

3 Na⁺

Na⁺

Na⁺ Na⁺ Na⁺

ATP

ATP
phosphorylates
α subunits

ADP

②

IN

P

Na⁺ Na⁺ Na⁺

P

dephosphorylation
stimulates
conformational
change

⑤

P

K⁺

K⁺ K⁺

K⁺

K⁺ K⁺

③

a conformational change
following phosphorylation
expels three Na⁺ into surroundings

3 Na⁺

④

two K⁺
accepted
from outside

2 K⁺

P

pump open to outside,
ready to start second
half of cycle

(b)

aspartate residue
(in α polypeptide
chain of the Na⁺/K⁺
ATPase)

ATP    ADP

phosphorylated
aspartate residue

(c)

***Figure 3.7*** (b) Schematic model of the
sodium/potassium pump in operation.
Steps of the pump's operations are shown
as a cycle. (c) The phosphorylation
reaction of Na⁺/K⁺ ATPase.

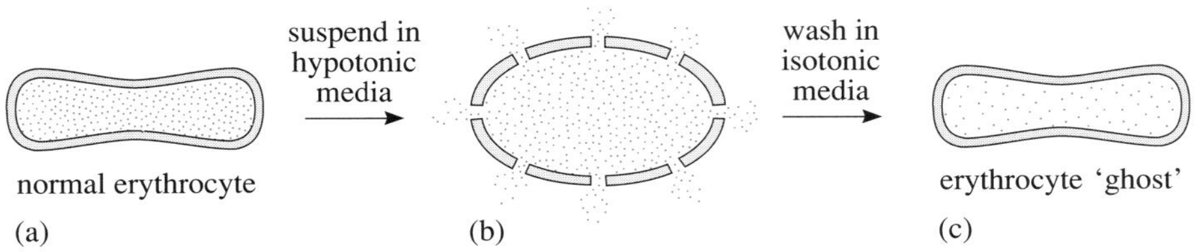

**Figure 3.8** The formation of erythrocyte ghosts. (a) An erythrocyte containing haemoglobin, in isotonic solution. (b) Cell swelling occurs in the hypotonic medium, causing the membrane to become leaky and so haemoglobin is lost and replaced by hypotonic solution. (c) After washing in isotonic media the erythrocyte reseals, with a new internal isotonic composition and a pale appearance.

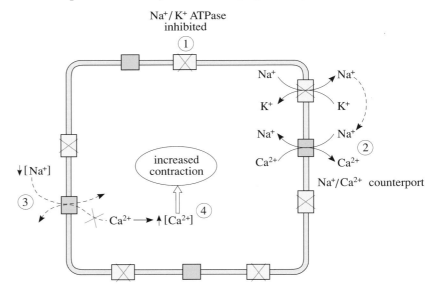

There are drugs that inhibit the dephosphorylation reaction of ATP by interacting with a specific binding site on the protein pump which is located on the extracellular face of the membrane. The clinical benefits of digitalis (which is derived from the foxglove plant) have been long recognized in the treatment of congestive heart failure: it inhibits the $Na^+/K^+$ pump, thus diminishing the $Na^+$ gradient which, in turn, leads to an increase in intracellular $Ca^{2+}$ concentration, since $Ca^{2+}$ is usually extruded by a $Na^+/Ca^{2+}$ counterport (Chapter 2, Section 2.4). The elevated cytosolic $Ca^{2+}$ concentration enhances the contractility of heart muscle and improves cardiac performance (Figure 3.9).

**Figure 3.9** The effect of digitalis on the heart. (1) The $Na^+/K^+$ ATPase is inhibited, diminishing the $Na^+$ gradient across the cell membrane. (2) $Ca^{2+}$ extrusion is coupled to $Na^+$ entry by a specific counterport. (3) The decreased $Na^+$ gradient leads to a decrease in the extrusion of $Ca^{2+}$ from the cell. (4) The increased intracellular $Ca^{2+}$ concentration enhances contractility of heart muscle cells.

### 3.5.2 Calcium ATPase

You will recall from Table 3.1 that the calcium concentration in the cytosol of resting (i.e. unstimulated) cells is very low ($0.1 \, \mu mol \, l^{-1}$) compared to that in the external medium ($1–2 \, mmol \, l^{-1}$). Electrical or chemical stimulation of cells may cause an increase (by up to 100-fold) in cytosolic $Ca^{2+}$ concentration and this serves as a mechanism for triggering numerous intracellular events such as muscle contraction, the release of hormones and neurotransmitters, the oxidation of metabolic fuels and cell division. The low intracellular $Ca^{2+}$ concentration in unstimulated conditions is maintained by a $Ca^{2+}$ ATPase located in the cell membrane and also in the endoplasmic reticulum (the lumen, i.e. the interior, of which constitutes a quantitatively important $Ca^{2+}$ store).

In skeletal muscle, the endoplasmic reticulum is a specialized network of tubules, termed the **sarcoplasmic reticulum** (Figure 3.10). Its function is to continually take up and release calcium ions which mediate contraction. The sarcoplasmic reticulum membranes are rich in $Ca^{2+}$ ATPase and have proved to be a good source of the protein for purification (it constitutes more than 80% of the integral membrane protein). DNA cloning and sequencing have revealed that the $Ca^{2+}$ ATPase has an amino acid sequence homologous to the $\alpha$ subunit of the $Na^+/K^+$ ATPase, suggesting that it functions in a similar manner, via the formation of a phosphorylated aspartate residue. The proposed phosphorylation and dephosphorylation cycle would induce the appropriate conformational changes which lead to the electrogenic extrusion of two $Ca^{2+}$ ions and the simultaneous hydrolysis of one molecule of ATP.

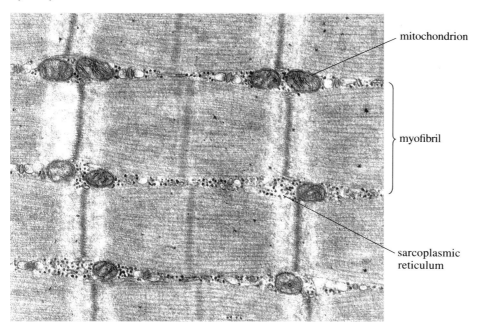

*Figure 3.10* Sarcoplasmic reticulum in muscle. This is a network of fine channels (here seen in cross-section) which serve as the calcium reservoir controlling muscle contraction. The myofibrils are made up of the contractile proteins myosin and actin. Note the numerous mitochondria between adjacent myofibrils.

### 3.5.3 Glucose transporters

At least five closely related proteins in the cell membrane are involved in the facilitated entry of glucose down a concentration gradient into mammalian cells. Glucose is able to passively permeate into all cells in the body due to the presence of one or more members of this transporter family in the cell membrane. These proteins exhibit the saturation and turnover kinetics of a typical carrier and show significant selectivity: for instance, the optical isomer of the naturally occurring D-glucose, L-glucose, is unable to enter cells via these transporters (see Table 3.2).

▷ What does the $K_M$ represent? Remember that the kinetics of transport are similar to those of an enzyme–substrate interaction (discussed in Book 2).

▶ $K_M$ refers to the extracellular concentration of sugar that is required to produce a rate of transport that is half the maximum (i.e. when saturation occurs). Thus, a transporter with a *high* affinity for its substrate will have a *low* $K_M$ value, e.g. D-glucose in Table 3.2.

*Table 3.2* The affinity* $K_M$ of the erythrocyte glucose transporter (GluT1) for different hexose sugars.

| Sugar | $K_M$/mmol $l^{-1}$ |
| --- | --- |
| D-glucose | 4 |
| D-galactose | 30 |
| D-xylose | 60 |
| L-galactose | >3 000 |
| L-glucose | >3 000 |

* May be referred to as $K_t$ in some texts.

The tissue distribution of the facilitated glucose transporter family is outlined in Table 3.3.

*Table 3.3* The transporters that mediate the facilitated entry of glucose into cells.

| Name of transporter | Tissue distribution |
| --- | --- |
| GluT1 | erythrocytes, endothelial cells lining the blood–brain barrier |
| GluT2 | liver, β cells* in the pancreas |
| GluT3 | neurons |
| GluT4† | skeletal muscle, heart, adipose tissue (fat) |
| GluT5 | small intestine, kidney |

* Also referred to as B cells in some texts.   † Insulin-sensitive.

The glucose transporter **GluT4** has been the subject of much investigation in recent years, following its identification as an insulin-sensitive glucose transporter. It is found in the cell membrane of two important insulin target tissues, muscle and adipose tissue, but only when insulin is present. In the absence of the hormone, GluT4 is internalized and relocated to intracellular vesicles which lie beneath the cell membrane (Figure 3.11a). Hence the mechanism of insulin-stimulated glucose transport involves the translocation of GluT4 to the cell surface, resulting in an increase in the number of transporters present and therefore a stimulation in the rate of entry of glucose into cells.

The active process of absorption of glucose into cells, particularly in the gut and kidneys (since the active uptake of glucose is an important function of these tissues), is achieved by a different family of transporters. The most common mode of accumulating glucose against a concentration gradient is by coupling its transport to the movement of $Na^+$ ions down a concentration gradient (Figure 3.11c).

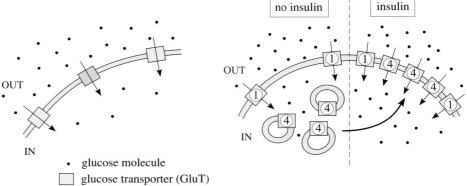

**Figure 3.11** Different modes of glucose transport. (a) Facilitated (carrier-mediated) diffusion. GluTs 1–5 present in the cell membrane promote the transport of glucose down its concentration gradient into the cell. (b) The regulation of GluT4 by insulin. In the absence of insulin, GluT1 present in the cell membrane allows basal transport of glucose; GluT4 is present in intracellular vesicles beneath the cell membrane. In the presence of insulin, Glut4 vesicles translocate to the cell membrane and fuse with it so increasing the rate of glucose transport. (c) Active transport by an $Na^+$-linked coport system. $Na^+$ moves down its electrochemical gradient coupled to the movement of glucose, which accumulates inside the cell. ATP is the ultimate energy source so this is *secondary* active transport.

(a) facilitated (carrier-mediated) diffusion

- glucose molecule
- □ glucose transporter (GluT)

(b) regulation of GluT4 by insulin (transporter translocation)

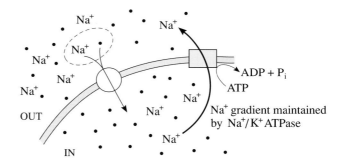

(c) active transport – $Na^+$-linked glucose coport system

▷ What is the term for this dual transport system and how, ultimately, is the energy provided?

▶ An $Na^+$-linked coport (or symport) drives the active uptake of glucose into cells. Energy is ultimately provide by ATP, since this molecule fuels the $Na^+$/$K^+$ ATPase which creates the $Na^+$ gradient; hence glucose enters using a secondary active transport system.

### 3.5.4  $HCO_3^-$/$Cl^-$ counterport

Erythrocytes contain an anion exchange protein which plays a critical role in transporting $CO_2$ from respiring tissues to the lungs for excretion. The protein, which exists as a dimer in the cell membrane, provides a channel for the tightly regulated exchange of one $Cl^-$ for one $HCO_3^-$ between intracellular and extracellular compartments, thus maintaining electrical neutrality (Figure 3.12). However, the way in which such a precise, stoichiometric counterport (i.e. 1 : 1 exchange) of the two anions is achieved has yet to be elucidated. In the lungs, $HCO_3^-$ leaves the erythrocyte and $Cl^-$ enters to compensate. $CO_2$ is formed from the released $HCO_3^-$ and this is then exhaled. In the tissues, where $CO_2$ is produced during respiration, the reverse exchange process occurs: $CO_2$ is converted into $HCO_3^-$ which in turn enters the erythrocyte in exchange for $Cl^-$. Thus the continual passive two-way shift of chloride ions according to the prevailing concentration of $CO_2$ provides a quantitatively important route for the transfer of $CO_2$ from the tissues to the lungs and thereby its excretion from the body.

### 3.5.5  ABC Transporters (traffic ATPases)

ABC transporters or 'traffic ATPases' are a family of transport proteins that are widely distributed throughout prokaryotic and eukaryotic organisms. (ABC stands for 'ATP binding cassette').  Members of this family exhibit a diverse range of functions. In bacteria they play a role in the import of small molecules, including peptides, sugars, amino acids and ions and also in the export of larger peptides. In humans, a chloride channel has been identified as a member of the ABC transporter family. There has been considerable interest in the function of this protein because it is genetically defective in cystic fibrosis, a disease characterized by defective electrolyte (ion) transport in several types of epithelia. It is thought that the resultant abnormalities in transepithelial chloride movement due to the ineffective channel protein – termed the cystic fibrosis transmembrane conductance regulator (CFTR) – account for many of the manifestations of the disease, particularly the abnormal accumulation of mucus in the lungs. Other physiologically important ABC transporters have subsequently been found in humans. These include TAP1 and TAP2, proteins that facilitate the entry of peptides into the endoplasmic reticulum and thereby regulate the process of antigen presentation in the immune system (see Chapter 4). Another example is P glycoprotein, which is involved in the active extrusion of hydrophobic chemotherapy agents from cells. By ridding cancer cells of potentially toxic drugs, P glycoprotein protects cancer cells from effective attack by chemotherapy agents (often termed 'multidrug resistance').

ABC transporter proteins all have a similar structural organization: they contain two hydrophobic membrane-spanning regions and two ATP binding domains (Figure 3.13). CFTR and P glycoprotein have hybrid properties, since they are able to operate as *either* transporters *or* channel proteins and can switch between these two configurations. ATP is an important accessory component for both forms of the transporter. ATP binding (but *not* hydrolysis) is obligatory for channel function, while the active carrier-mediated extrusion of drugs from the cytoplasm requires an energy source, namely ATP hydrolysis.

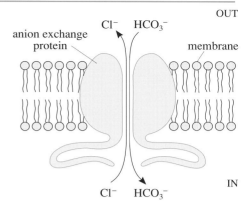

**Figure 3.12**  The ion transport protein of erthrocyte membranes. This integral membrane protein exists as a dimer in the membrane. It serves to exchange $Cl^-$ and $HCO_3^-$ ions across this membrane.

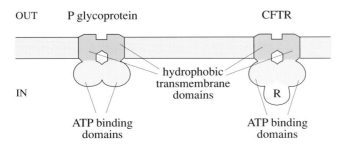

**Figure 3.13**  The structural organization of two important ABC transporters in humans. Both transporters are single polypeptide chains, composed of two membrane-spanning regions and two ATP binding domains. CFTR has a fifth domain which serves a regulatory role (R).

## 3.6  Bulk transport

There are two other important modes of transport across the cell membrane which should also be considered.

The intake of both particulate matter and the soluble contents of extracellular fluid is facilitated by **endocytosis**. During this process, the cell membrane encloses a region of extracellular space by invagination, then pinches off to form an intracellular vesicle containing the ingested material (Figure 3.14a). Two types of endocytosis are distinguished on the basis of the size of vesicles formed. **Pinocytosis** (which means 'cell drinking') involves the internalization of small fluid-filled vesicles containing solutes. **Phagocytosis** (which means 'cell eating') refers to the uptake of solid particulate matter such as macromolecules and cell debris into larger vesicles, where digestion ensues.

**Figure 3.14** (a) Endocytosis and (b) exocytosis. The extracellular space is at the top, separated from the cytosol below by the cell membrane. Note that endocytosis and exocytosis are not simply the reverse of each other; in exocytosis two cytosolic-side monolayers of the cell membrane initially adhere, whereas in endocytosis two non-cytosolic-side monolayers initially adhere. In both cases the asymmetric character of the membranes is conserved and the cytosolic-side monolayer remains in contact with the cytosol.

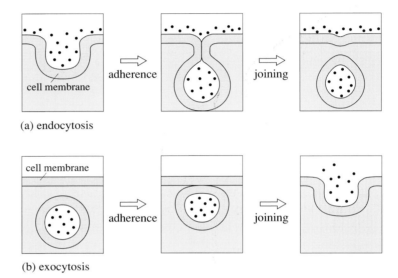

(a) endocytosis

(b) exocytosis

Most cells employ pinocytosis as a method for the continual ingestion of small volumes of fluid and solutes across the cell membrane. In contrast, the uptake of particulate matter is a function restricted to specialist cells. The champion of phagocytosis is a type of immune cell called a **macrophage**, which is able to engulf 25% of its volume in one hour (Figure 3.15). Since its volume and surface area are not significantly altered during this process, membrane must be added at a rate which matches its removal from the surface: this is achieved by the recycling of phospholipid bilayer back to the cell membrane following the internalization process.

Many endocytotic vesicles ultimately fuse with lysosomes (described in Book 1, Chapter 1). As lysosomes house a spectrum of degradative enzymes, the resultant vesicle becomes a specialized site for intracellular digestion. Breakdown products, such as sugars, amino acids and nucleotides, are then available for cell use. However, some vesicles bypass the lysosome pathway and instead act as a shuttle system, shifting material from one cell surface to another. This is considered to be an important function of endocytosis in endothelial cells, where it serves as a one means of translocating substances from the blood into the extracellular space.

A specialized mode of endocytosis employs surface receptors that allow the selective uptake of substances into cells. This so-called **receptor-mediated endocytosis** (RME) is a rapid internalization process triggered by receptor occupancy.

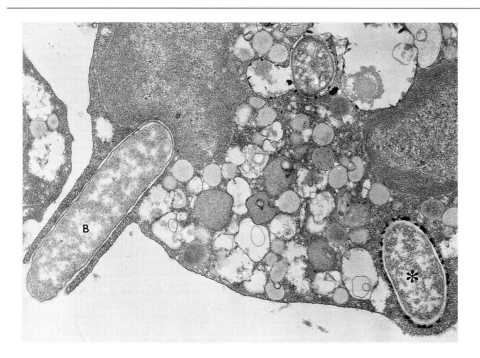

***Figure 3.15*** Electron micrograph showing phagocytosis of a bacterium by a macrophage. Note the protrusion of the cell membrane around the bacterium (B) and another bacterium enclosed within the cell (*). ($\times 20\,000$)

The bulk transport of material *out* of the cell is facilitated by **exocytosis**. This is functionally the reverse of endocytosis, whereby intracellular vesicles fuse with the cell membrane and disgorge their contents into the extracellular space (but see Figure 3.14b). Exocytosis is the exit route for many secreted substances; for example, plasma proteins, hormones and neurotransmitters (e.g. adrenalin and acetylcholine) are expelled from the cell by this process. Exocytosis also facilitates the transport of phospholipid to the cell membrane.

Some cells release intracellular vesicles continuously; for example, this is the route by which liver cells release albumin and plasma proteins into the bloodstream. Alternatively, the exocytosis of many substances is triggered by a rise in intracellular $Ca^{2+}$ concentration. The molecular mechanism is thought to involve cytoskeletal proteins and plays a particularly important role in intracellular signalling, as it controls the processes of neurotransmission at synapses and the secretion of hormones.

### 3.6.1  Receptor-mediated endocytosis

Receptor-mediated endocytosis is an important route for the selective uptake of substances into the cell. It is also important for regulating the sensitivity of cells to external stimuli.

The process involves the clustering of receptors into specialized regions of the cell membrane called *coated pits*. These have a mesh-like lining on the cytosolic face of the bilayer due to the presence of a protein called *clathrin*. The binding of a ligand to a receptor located in a coated pit region of the membrane initiates rapid invagination which generates coated intracellular vesicles. These then release the clathrin, forming a smooth-coated *endosome* (Figure 3.16) which can undergo fusion with lysosomes leading to the degradation of ligand; the receptors themselves are often preserved and recycled back to the cell surface (Figure 3.16). For this to occur, the ligand–receptor complex must dissociate and this event is promoted by the lowering of the pH inside the vesicle. Most endosome membranes contain ATP-dependent proton pumps which generate the required intravesicular acidity.

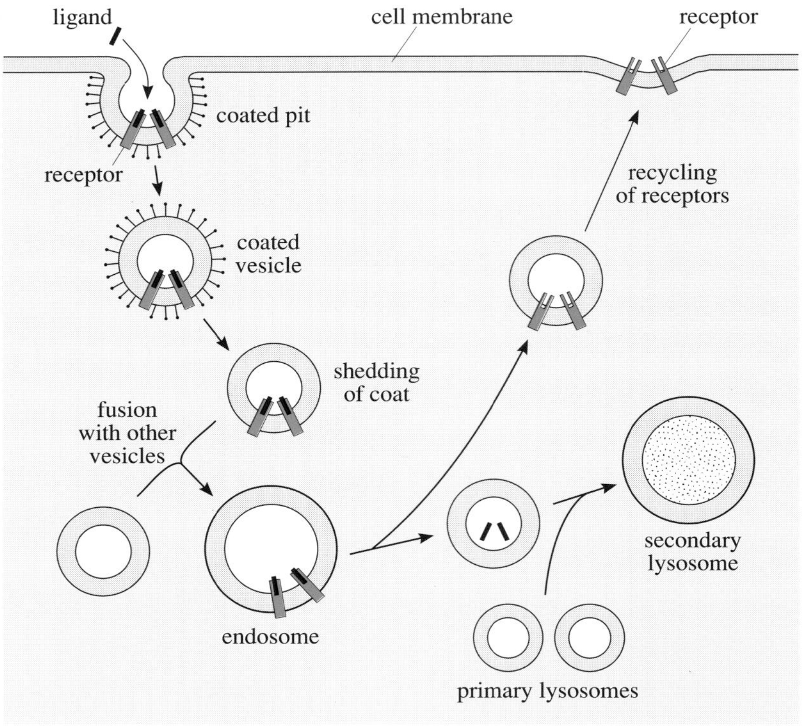

**Figure 3.16** Receptor-mediated endocytosis. Receptor–ligand complexes are endocytosed from the cell membrane in a coated vesicle. Note that while the ligands are broken down in a secondary lysosome, the receptor proteins are retrieved from the endosome and recycled to the cell membrane. This scheme shows just one possible result of endocytosis; degradation of receptors in lysosomes is also possible. There must therefore be mechanisms regulating such intracellular 'membrane traffic'.

### Receptor-mediated endocytosis and the uptake of cholesterol

The RME process was elucidated by Michael Brown and Joseph Goldstein in the 1970s at the University of Texas, during their investigations on the nature of cholesterol uptake by cells. Cholesterol is transported in the bloodstream in the centre of a phospholipid-bound particle called low-density lipoprotein (LDL). Receptors for LDL are present in the coated pit regions of the cell membrane of many cells and these interact with a specific protein on the surface of LDL particles. As expected, the binding of LDL to its receptor promotes rapid internalization, followed by endosomal degradation of the particle as illustrated in Figure 3.16. Cholesterol is then liberated for use in the cell.

LDL receptor-mediated endocytosis regulates the concentration of cholesterol in the blood. A defective RME pathway has been identified in the hereditary condition called familial hypercholesterolaemia. Individuals with this genetic abnormality have extremely high levels of cholesterol in their blood and this predisposes them to premature atherosclerosis ('hardening of the arteries') and death from coronary heart disease. Brown and Goldstein established that the impairment is due to a failure of the receptor protein involved in LDL uptake. A number of mutant LDL receptors have now been identified: these are either unable to bind LDL or unable to become localized in the coated pit regions of the cell membrane.

### Other functions of receptor-mediated endocytosis

RME is one means of regulating the number of receptors on the surface of a cell. In some cases, receptor phosphorylation promotes its internalization (discussed further in Section 3.8.3), perhaps by stabilizing the receptor protein in the region of a coated pit. This is one of the major mechanisms of *desensitization* – the process that results in a loss of response by a cell. When cells are exposed to high concentrations of a ligand, a decrease in the number of receptors present on the cell membrane is often observed. In this way, RME may contribute to the desensitization observed in insulin-resistant diabetic patients (see Chapter 4).

Receptor-mediated endocytosis also plays a number of critical roles in the fight against infection. For example, the immune system's own specialized phagocyte, the macrophage, expresses receptors that are able to bind and recognize commonly encountered microbial structures, and these promote the removal and destruction of invading pathogens. These receptors also play a role in intracellular signalling and stimulate a number of other processes in the macrophage. Receptor-mediated endocytosis is also important in a more elaborate system of immunological recognition involving a class of white blood cells known as lymphocytes.

## Summary of Sections 3.2–3.6

The cell membrane must have mechanisms which allow the entry and exit of metabolites and waste products and maintain cell responsiveness to the environment (discussed further in the remainder of this chapter and in Chapter 5). Non-polar substances such as $O_2$ and $CO_2$ readily cross the phospholipid bilayer and so enter and leave cells by simple diffusion. However, polar compounds cannot penetrate the central lipid core of the cell membrane and so these substances traverse the bilayer with the aid of specialized transport proteins. Such transport proteins are characterized by hydrophobic membrane-spanning sequences which favourably interact with the central lipid core and also provide a hydrophilic low-resistance route for charged and polar substances to cross the bilayer.

In the absence of an energy source, transport proteins facilitate the passive movement of uncharged solutes down a concentration gradient until the concentrations are equal on both sides of the membrane. Charged ions migrate according to both the concentration gradient and the voltage distribution across the cell membrane until equilibrium is reached (according to the Nernst equation – Chapter 2).

Substances may be transported against a prevailing concentration gradient (which is electrochemical in the case of charged ions) if a free energy source is supplied. Typically, ATP fuels primary transport systems; $Na^+/K^+$ ATPase is the most quantitatively significant pump in animal cells, responsible for setting up an asymmetric distribution of ions across the cell membrane. The electrochemical gradients of $Na^+$ and $K^+$ represent an energy store; one of the functions is to drive secondary coports and counterports. It also plays an important role in the conduction of nervous impulses and this aspect will be discussed further in Chapter 5.

The proteins that mediate both passive and active transport across the cell membrane may be described as channels and/or carriers. The distinction between the two is not always clear but is usually defined by the turnover number of the transporter. Generally, carriers exhibit high affinity associations with solutes and a correspondingly low turnover numbers. In contrast, movement through channels (which is always in the direction of the prevailing gradient) is considerably more rapid and often dependent on some form of regulation (or 'gating'). However, in most cases, channels display some degree of selectivity and this is considered to retard the rate of transport through the channel pore, thus leading to turnover numbers that are lower than you would expect for a typical channel. Channels are most commonly employed for the movement of ions across the cell membrane.

Finally, the cell membrane also employs the processes of endocytosis and exocytosis as a means of transport and communication between the cell and its exterior. Together the two mechanisms allow the bidirectional movement of both soluble and particulate matter and may also, on occasions, be precisely and specifically regulated. Cells employ receptor-mediated endocytosis (RME) for the selective uptake of substances such as cholesterol and for regulating the number of receptors at the cell membrane (desensitization). Internalization of receptors and their corresponding ligands occurs in specialized membrane regions called coated pits, which are lined with a protein (clathrin). RME has pathological significance in a number of clinical situations, e.g. in the inherited condition familial hypercholesterolaemia and in the desensitization found in insulin-resistant diabetic patients (see Chapter 4).

## 3.7   Intracellular communication: responses to external stimuli

### 3.7.1   Introduction

The cell membrane isolates a cell from its environment but, as you have just discovered, a variety of specialized transport systems permit the movement of chemical traffic between the cell and the surrounding media and also maintain a cell's unique ionic composition. The phospholipid bilayer is, however, more than just a selective permeability barrier; it is intimately involved in the transmission of information from the external environment, allowing the cell to monitor and respond to the changing conditions around it. Information from external stimuli may regulate numerous physiological processes such as nutrient uptake, cell proliferation and differentiation. It is the intracellular signalling process which is the focus for the remainder of this chapter.

In the case of multicellular organisms, the cell membrane is also important in biocommunication between cells, many of which perform distinct functions. This intercellular form of communication (distinct from intracellular signalling, as illustrated in Figure 3.17) is the subject of Chapter 4.

### 3.7.2   Signal processing by cells

Cells respond to a variety of external stimuli using a limited repertoire of intracellular signalling pathways. The term **transduction** is often used to describe the biochemical processes that operate when information is relayed from the environment to the inside of the cell, leading to the generation of specific biological responses. Such signal processing pathways involve a number of components through which information flows in a set sequence. Figure 3.18 indicates the major constituents of these pathways and outlines the route for the transduction of a stimulus into the cell.

The most common external stimulus is chemical; this category includes agents that stimulate the senses, such as taste and smell, along with physiological mediators such as hormones, neurotransmitters and cell surface molecules – all of which you will encounter in the next chapter. In addition, cells are often capable of responding to other factors, such as light, voltage change and stretch/mechanical distortion. In the forthcoming sections we shall look at a number of specific examples of the different routes employed by cells as a means of intracellular signalling, but before we launch into detail let us first consider some of the basic principles of transduction:

(a) intracellular communication

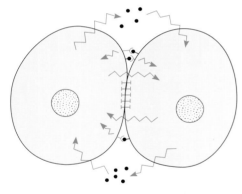

(b) intercellular communication

- receptor–ligand interaction
- gap junction
- chemical mediator

*Figure 3.17*   Intracellular and intercellular communication. (a) Intracellular communication: external stimuli (e.g. chemicals, light, movement) are detected by a cell, activating signalling pathways inside the cell (in various intracellular compartments). (b) Intercellular communication: information is exchanged between cells either by physical contact or by chemical mediators (discussed in detail in Chapter 4).

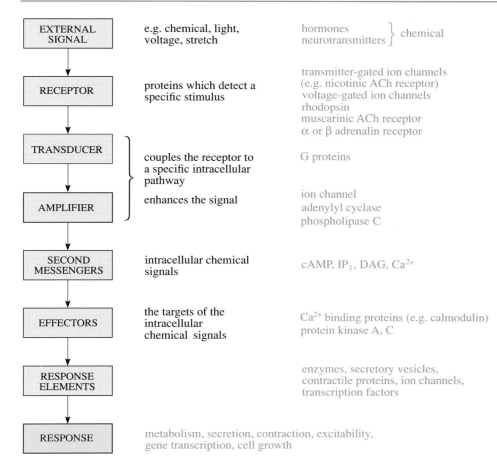

**Figure 3.18** The major components of signal processing pathways and the route of the transduction cascade from stimulus to response. The examples given here are ones you will encouter in this chapter. (*Note*: a given pathway need not involve every component.) IP3 = inositol trisphosphate; DAG = diacylglycerol; ACh = acetylcholine.

1 Information is passed from one component to the next using two basic mechanisms: conformational coupling or covalent modification (Figure 3.19).

2 A hallmark of signal processing cascades is their ability to amplify the initial signal. A single hormone–receptor interaction can be enhanced up to a million times. Although shown at just one stage in Figure 3.18, amplification can occur at a number of steps in the pathway.

3 Different signalling pathways often coexist in cells, and there is frequent overlap of the response elements targeted. This redundancy means that alternative pathways can compensate if any one signalling system is lost.

4 Signal termination is a key feature of all signal transduction processes. Some kind of 'switch-off' system is essential if a cell is to maintain its sensitivity to

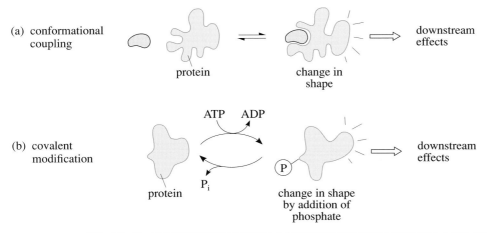

**Figure 3.19** The transfer of information in signalling pathways. (a) Conformational coupling: non-covalent interactions between components induce conformational changes in proteins (analogous to the allosteric regulation of enzymes – see Book 1, Chapter 2), which in turn have effects on downstream components. (b) Covalent modification: the addition or removal of a phosphate group onto or from a protein affects its activity and function, including its effects on downstream components.

external stimuli. A variety of termination mechanisms operate and these can be modulated to alter the magnitude and duration of the response.

The biological responses to external stimuli may be observed over a broad time-span. Some are immediate, usually due to the rapid activation or inhibition of ion fluxes, cellular enzymes and/or transport proteins. Others may be seen hours or even days after the initial stimulus. These long-term effects are usually attributed to alterations in the pattern of gene expression within responding cells and often associated with cell growth and development.

### 3.7.3 Receptors: their role in signal transduction

Most intracellular signalling pathways involve the detection of a stimulus by a receptor at the cell membrane. Indeed, the cell membrane has a prominent role in transduction and signal amplification, as you will discover later. However, you should be aware that not all intracellular transduction systems involve cell membrane receptors. One class of chemical mediators – those that are lipophilic (such as steroid hormones, which are discussed in Chapter 4) – readily diffuse through a lipid bilayer and interact reversibly with specific intracellular receptor proteins. A steroid receptor complex then undergoes a conformational change which allows it to bind to DNA in the nucleus of the cell, switching on the transcription of primary response ('immediate early') genes. In turn, the products of these early genes then act as signals to activate others (see Chapter 1, Section 1.4, and Figure 3.20). In the majority of cases there is a considerable time lag between receptor activation and the observed cellular response, namely an alteration in the pattern of gene expression. This is because the gene regulation cascade involves transcription and translation and these events may take hours or longer.

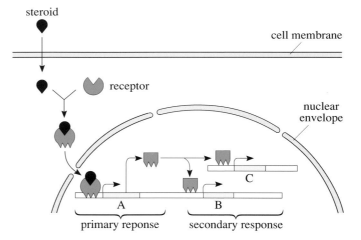

*Figure 3.20* Signal transduction through intracellular steroid receptors. Steroid hormones diffuse across the cell membrane of the target cell and bind reversibly to specific steroid hormone receptors in the cytosol. The steroid–receptor complex then undergoes a conformational change which allows it to bind to DNA to switch on the transcription of primary response or 'immediate early' genes (A) which produce signals to activate other genes (secondary response, B and C).

### 3.7.4 Cell membrane-bound receptors

Having considered intracellular receptors and their associated signalling cascades, we shall now turn our attention to some specific examples of receptors located at the cell membrane. We shall look at how they detect stimuli and how they relay information to downstream components.

For convenience, this very large group of receptors can be further subdivided as follows:

1  Receptor ion channels
2  Receptors that are coupled to G proteins
3  Receptors that also have intrinsic enzyme activity

The groupings refer not only to differences in structure, but also in their modes of signalling, which we shall now consider in further detail.

### Receptor ion channels

These multisubunit transmembrane protein complexes are both receptors *and* ion channels, the opening of which is controlled (or gated) by a specific external stimulus. Members of this class of receptors have been found to be regulated by stretch, voltage or, most commonly, chemical signals. In all cases, the stimulus induces a conformational change in the receptor which may either open or close the ion channel, thereby altering ion fluxes into or out of the cell. Signalling through this type of receptor is extremely rapid: many of the components of the transduction pathway shown in Figure 3.18 may be 'short-circuited', so that cellular responses may be generated almost immediately. (This is in sharp contrast to the steroid receptor signalling pathway, where responses take much longer to develop.) Gated ion channels usually have self-closing mechanisms, even if the stimulus which originally opened them is still operating. This is an important means of signal termination.

### (a) Transmitter-gated ion channels

The term **transmitter-gated** ion channel refers specifically to receptor ion channels that are gated by chemical substances. (Chemical substances that bind to and activate receptors are often called *ligands*, which is why these receptors are also referred to as *ligand-gated* ion channels – Chapter 2, Section 2.5.1.)

In humans, the most well-characterized transmitter-gated ion channels are found in muscle, at the junction where incoming nerves converge, providing a signal for contraction. Electrical impulses travel down these nerves and stimulate the release of acetylcholine, a chemical mediator (termed a **neurotransmitter**, and discussed in more detail in Chapter 4) into the synaptic cleft between the nerve endings and muscle cells (Figure 3.21). Like many neurotransmitters, acetylcholine is recognized by a number of different receptors. The different types of acetylcholine receptors can be distinguished in pharmacological studies, since each type binds a different spectrum of acetylcholine-like molecules. For example, the receptors at the neuromuscular junction bind not only acetylcholine but also nicotine, and are hence called **nicotinic** receptors. The ability to respond to nicotine distinguishes this class of receptor from a different type of acetylcholine receptor, the **muscarinic** receptor, which also responds to muscarine, a substance derived from mushrooms, and is present mainly in the brain. Although both nicotine and muscarine are ligands which bind to different types of acetylcholine receptors, it is important that you recognize that neither plays a physiological role in the regulation of the receptor in normal life.

Nicotinic receptors have been the focus for many studies because they were found to be present at very high concentrations in the electric organs of fish (particularly electric eels – Plate 2.1) and these organs subsequently proved to be a rich source of the protein for isolation and purification. The nicotinic receptor proteins of fish electric organs can be visualized using electron microscopy (Figure 3.22). Note their ring-like appearance, along with a central cavity which takes up an aqueous stain.

▷ What do you think this cavity represents?

▶ It is the pore of the ion channel.

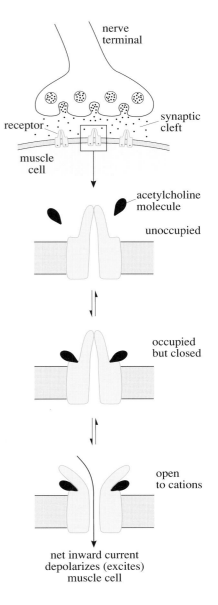

**Figure 3.21** The response of the acetylcholine receptor to acetylcholine. (Prolonged exposure to high concentrations of acetylcholine causes the receptor to enter yet another state – not shown – in which it is inactivated and will not open even though acetylcholine is present.)

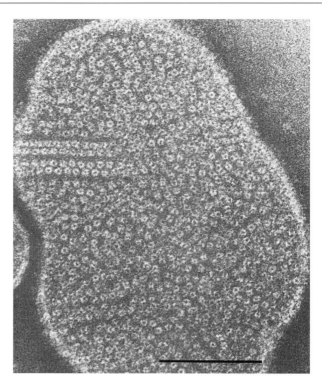

**Figure 3.22** The nicotinic receptor proteins found in the membranes of the *Torpedo* electric organ, visualized by electron microscopy (scale bar = 0.1 μm). See text for details.

The nicotinic receptor is a pentamer (i.e. made up of five subunits), consisting of two α polypeptides in addition to one each of β, γ and δ (Figure 3.23). Adjacent receptors are further linked to form dimers of the five-subunit structure.

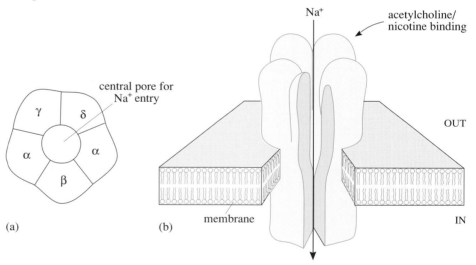

**Figure 3.23** The structure of the nicotinic acid receptor. (a) View from above, looking at the extracellular face. (b) Cross-section, showing the receptor protruding from the membrane on both sides.

▷ How might the four components of the nicotinic receptor (α, β, γ and δ) be separated into distinct polypeptides?

► Denaturation of the purified protein complex (by a detergent such as sodium dodecyl sulphate, SDS) disrupts the inter- and intramolecular protein interactions. Electrophoresis is then used to separate the different protein components on the basis of their size and charge (Book 1, Chapter 3).

The nicotinic receptor channel is cation-selective: $Na^+$ is the major ion that traverses the bilayer via the receptor's protein pore, migrating down its concentration gradient into the cell when the channel is open (Figure 3.23). The binding of acetylcholine (or nicotine) is presumed to induce conformational changes in the structural organization of the receptor. This is the trigger for the opening of the channel and an increase in the permeability of the cell to Na$^+$, and this results in depolarization of the cell. Acetylcholine is thus an **excitatory** neurotransmitter

Another important transmitter-gated ion channel present in the brain is the anion-selective GABA (gamma-aminobutyric acid) receptor. This complex, which like the nicotinic receptor is pentameric, functions as a ligand-regulated Cl$^-$ channel and in the presence of GABA opens and lets chloride ions into the cell. The increase in permeability to chloride makes a cell's membrane potential more negative, rendering depolarization less likely. Because of this, GABA is described as an **inhibitory** neurotransmitter.

So far we have considered receptors that are gated by *extracellular* chemical stimuli. However, there are a number of transmitter-gated ion channels that are regulated by *intracellular* ligands (Table 3.4). All the receptors listed in Table 3.4 conform to the basic structural configuration of oligomeric proteins (i.e. they are made up of several subunits) which assemble in the bilayer to form a central membrane pore. Intracellular ligands are usually themselves generated by the activation of another class of receptors and thus serve as second messengers in the signal transduction cascade shown in Figure 3.18.

**Table 3.4** Some examples of transmitter-gated ion channels/receptors.

| Activating ligand | Major ion species which traverses the channel | Important tissues/organelles where channels are found |
|---|---|---|
| *Extracellular* | | |
| acetylcholine (nicotinic receptors) | Na$^+$ | nerves, muscle |
| GABA | Cl$^-$ | brain |
| glycine | Cl$^-$ | brain stem/spinal cord |
| *Intracellular* | | |
| cGMP | Na$^+$ | photoreceptor cells (eye) |
| cGMP, cAMP | Na$^+$ | olfactory neurons |
| inositol trisphosphate (IP$_3$) | Ca$^{2+}$ | endoplasmic reticulum |

## (b) Voltage-gated ion channels

You will be familiar with this group of receptor proteins from Chapter 2. Their channel structure undergoes a conformational change (becomes activated) in the presence of an electrical stimulus. This activation initiates an alteration in ion distribution (Figure 3.24), and therefore changes the voltage across the cell

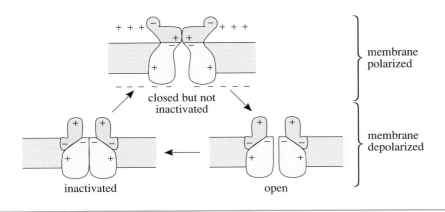

closed but not inactivated

membrane polarized

membrane depolarized

inactivated

open

**Figure 3.24** The regulation of voltage-gated ion channels. The altered ion distribution which occurs when the membrane is depolarized results in a conformational change in the ion channel, which opens allowing specific ions to enter or exit from the cell. This ion movement leads to rapid channel inactivation. The channel adopts its closed (but not inactive) state when the membrane is repolarized. It is now poised to respond to subsequent depolarizations.

membrane. Voltage-gated channels are present in all electrically active cells, where they are responsible for the propagation of depolarizing action potentials. You will learn more about electrically sensitive $Na^+$ and $K^+$ channels in Chapter 5, where their role in nervous conduction along axons is considered in more detail.

### Receptors that are coupled to G proteins

All the receptor proteins in this class are characterized by the presence of seven transmembrane α-helices. Once activated by the appropriate stimulus, they operate via an intermediary molecule, called a **G protein** (guanine nucleotide-binding protein), which in turn regulates the activity of enzymes or ion channels (the signal amplifiers in Figure 3.18) in the cell membrane. Many chemical stimuli bind and activate G protein-linked receptors. Such ligands include hormones such as adrenalin and agents that stimulate smell and taste receptors (Table 3.5). Interestingly, the neurotransmitter acetylcholine also signals via a G protein system, but in this case the muscarinic, not the nicotinic, receptor is involved.

*Table 3.5*  Some of the physiological effects mediated by G proteins.

| Stimulus | Cell type affected | G protein | Effector | Effect |
|---|---|---|---|---|
| adrenalin, glucagon | liver cells | $G_s$ | adenylyl cyclase | breakdown of glucogen |
| adrenalin, glucagon | fat cells | $G_s$ | adenylyl cyclase | breakdown of fat |
| luteinizing hormone | ovarian follicles | $G_s$ | adenylyl cyclase | increased synthesis of oestrogen and progesterone |
| antidiuretic hormone | kidney cells | $G_s$ | adenylyl cyclase | conservation of water by kidney |
| acetylcholine | heart muscle cells | $G_i$ | potassium channel | slowed heart rate and decreased force of contraction |
| enkephalins, endorphins, opioids | brain neurons | $G_i/G_o$ | $Ca^{2+}$ and $K^+$ channels, adenylyl cyclase | changed electrical activity of neurons |
| angiotensin | smooth muscle cells in blood vessels | $G_q$ | phospholipase C | muscle contraction, elevation of blood pressure |
| odorants | neuroepithelial cells in nose | $G_{olf}$ | adenylyl cyclase | detection of odorants |
| light | rod and cone cells in retina | $G_t$ | cyclic GMP phosphodiesterase | detection of visual signals |

The detection of light by the photosensitive rod cells of the retina is another G protein-mediated event. The receptor protein, called **rhodopsin**, contains an 11-*cis*-retinal light-absorbing group (*chromophore*) and this spontaneously isomerizes to the all-*trans* form on exposure to light. The energy of a light photon is thus converted into atomic motion due to the altered geometry of the all-*trans* species and the receptor is said to be activated. Specific details concerning the mode of action of G proteins are discussed in Section 3.8.

### Receptors that also have intrinsic enzyme activity

The proteins belonging to this class of receptors traverse the lipid bilayer of the cell membrane only once. They respond exclusively to chemical stimuli, but bind a diverse array of polypeptide ligands, including hormones (e.g. insulin, natriuretic peptides – see p.77), *cytokines* (chemical mediators of the immune system, discussed further in Chapter 4) and *mitogenic growth factors* (e.g. epidermal growth factor, platelet derived growth factor, insulin-like growth factor I (Figure 3.25).

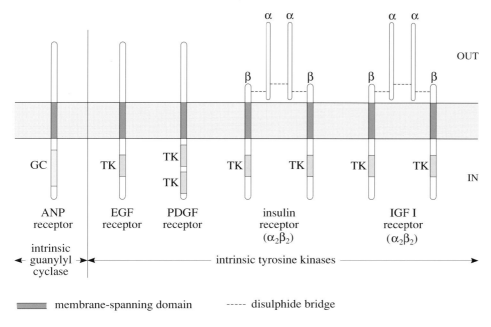

OUT

IN

GC    TK    TK    TK    TK    TK    TK
             TK

ANP      EGF      PDGF     insulin         IGF I
receptor receptor receptor receptor       receptor
                          (α₂β₂)          (α₂β₂)

intrinsic
guanylyl ← → intrinsic tyrosine kinases →
cyclase

▮ membrane-spanning domain    ----- disulphide bridge

**Figure 3.25** Receptors with intrinsic enzymic activity. Many of these have tyrosine kinase (TK) activity and some have guanylyl cyclase (GC) activity, both of which are within the cytosolic domain of the receptor protein. All are characterized by a single membrane-spanning domain. ANP, atrial natriuretic peptide; EGF, epidermal growth factor; PDGF, platelet-derived growth factor; IGF I, insulin-like growth factor I. Note that both EGF and PDGF receptors often exist as dimers in the membrane and that the insulin and IGF I receptors are each made up of four chains ($\alpha_2\beta_2$) covalently linked by disulphide bridges.

The natriuretic peptides are mammalian hormones that regulate salt balance. Their receptors fall into a distinct subtype which possess guanylyl cyclase activity (Figure 3.25, left). Hormone binding enhances the enzymic activity of the receptors, and increases the rate of synthesis of cGMP from GTP (Figure 3.26).

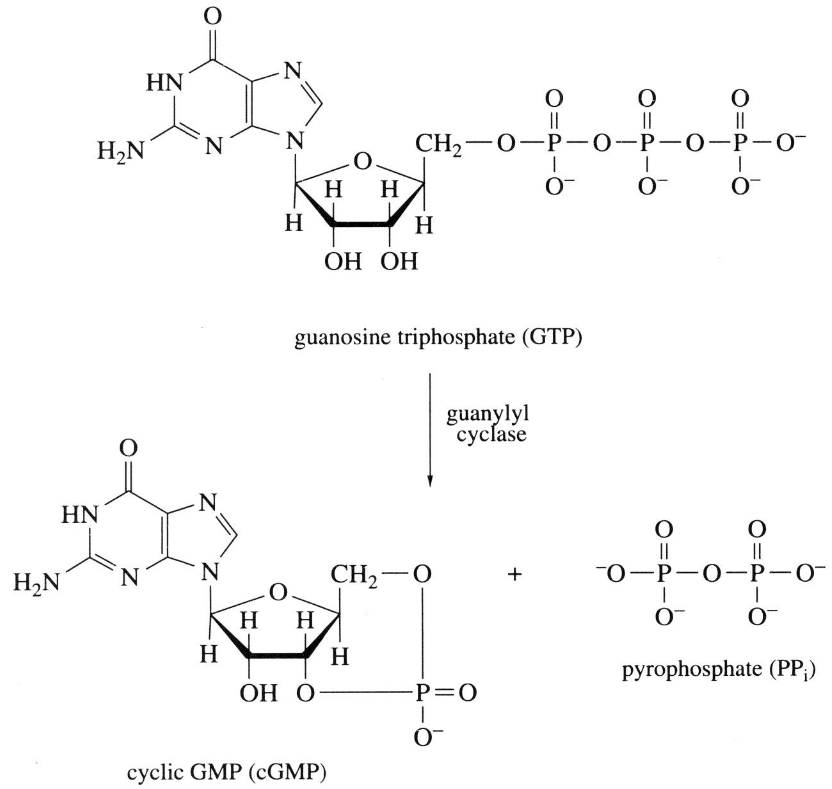

guanosine triphosphate (GTP)

guanylyl cyclase

cyclic GMP (cGMP)

pyrophosphate (PP$_i$)

**Figure 3.26** The guanylyl cyclase reaction.

▷ Can you recall one possible effect of an increase in intracellular cGMP levels? (Table 3.4 should help you.)

▶ cGMP acts as a second messenger, perhaps by regulating a specific transmitter-gated $Na^+$ channel.

Many of the receptors in this class have tyrosine kinase activity, the site of which is also located in an intracellular domain of the receptor protein (Figure 3.25). This enzymic function is, again, stimulated when appropriate ligands bind and activate the receptor.

All **kinase** enzymes catalyse phosphorylation reactions (Figure 3.27). They are characterized by the presence of a binding site for ATP.

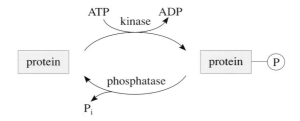

**Figure 3.27** The enzymes involved in protein phosphorylation and dephosphorylation.

▷ What is the role of the ATP in the kinase reaction?

▶ It functions as a coenzyme, donating a phosphate group to the reaction substrate.

Phosphate groups are often added to proteins, at specific amino acid residues on the polypeptide chain (Figure 3.28).

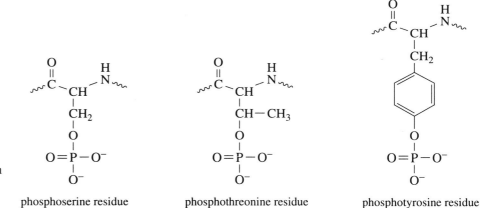

**Figure 3.28** The covelent modification of proteins by phosphorylation

phosphoserine residue          phosphothreonine residue          phosphotyrosine residue

▷ What do the three amino acid targets for kinases (tyrosine, threonine and serine) all have in common?

▶ They all possess an OH group, to which the phosphate is covalently attached.

In contrast, a **phosphatase** is an enzyme that reverses the effect of a kinase, i.e. it removes covalently-attached phosphate groups (Figure 3.27).

The substrates for receptor tyrosine kinase enzymes include the tyrosine residues present within the cytosolic face of the transmembrane polypeptide itself and the tyrosine residues of numerous other intracellular proteins, many of which have yet to be characterized. Some of the intracellular targets have been shown to be other kinases and phosphatases, resulting in the establishment of complex **phosphorylation signalling cascades**. Together, the opposing processes of phosphorylation and dephosphorylation regulate many intracellular processes within the cell, affecting the rate of solute transport and enzyme activity, and modifying the organization of cytoskeletal proteins and gene expression. Hence covalent modification (in the form of phosphorylation) is an extremely important process, central to the control of many biological responses.

However, signalling via the tyrosine kinase pathway is not necessarily the only route of signal transduction from the receptor into the cell. For instance, the insulin receptor employs a G protein as an additional means of intracellular signalling. Since many of the tyrosine kinase receptors control a wide spectrum of biological effects (particularly the complex processes of cell growth and proliferation), a number of distinct modes of transduction are likely to be recruited.

### 3.7.5  Receptor pharmacology

The binding of chemical stimuli, such as hormones and neurotransmitters, to receptors usually occurs with a high degree of specificity.

▷ The types of interactions that stabilize the ligand–receptor complex parallel those of an enzyme and its substrate (discussed in Book 2) and also those of a transporter and its solute, mentioned briefly earlier in this chapter. What sorts of molecular forces are involved?

► Hydrogen bonds and electrostatic, hydrophobic and van der Waals interactions.

There is often more than one type of receptor for a given ligand.

▷ What is the example you have already encountered?

► Acetylcholine binds to nicotinic and muscarinic receptors.

Different receptors may employ distinct signalling mechanisms.

▷ How do nicotinic and muscarinic receptors differ in this way?

► The nicotinic receptor is also an ion channel which opens in response to ligand binding. The muscarinic receptor is coupled to a G protein.

The hormone adrenalin also signals through two different receptor proteins (called α- and β-receptors). In this case, both are coupled to G proteins but, as you will learn later, each class of receptor activates a distinct type of G protein, which results in different biological responses inside cells.

The magnitude of a biological response to an external stimulus is proportional to its size (for example, the extracellular concentration of a ligand), until a maximal response is achieved (Figure 3.29).

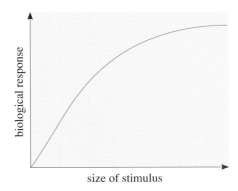

*Figure 3.29*  The relationship between size of stimulus and magnitude of the biological response.

Chemical compounds similar in structure to natural ligands (such as acetylcholine or adrenalin) may also interact with receptor proteins. This aspect of receptor biology has provided an opportunity for the development of many therapeutic drugs. Typically, there are two possible outcomes of the interaction of such compounds with a receptor protein. The compound may mimic the biological effect of the ligand, in which case it is said to behave as an **agonist**. Nicotine and muscarine are such agonists, although they are of little clinical relevance. However, the β-receptor agonist salbutamol (often referred to as Ventolin) is a very important drug used in the treatment of asthma. If it is inhaled during an attack it promotes the relaxation of the lung bronchioles via its activation of β-receptors, thus relieving the distressing effect of airway constriction.

Compounds that bind to receptors but fail to activate the subsequent signalling pathway block the effect of the natural ligand by preventing its access to the binding site. Such agents are described as **antagonists**. Propanolol and other so-called 'β-blockers' are antagonists, and are used to limit the stimulatory effects of adrenalin on the heart. They are of benefit to patients suffering from high blood pressure and cardiac arrhythmias. The effects of adrenalin and propanolol on the heart rate response are shown in Figure 3.30.

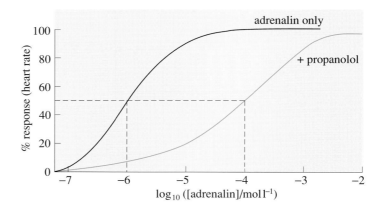

**Figure 3.30** Effect of propanolol (1 μmol l$^{-1}$) on the response of heart rate to adrenalin.

The affinity of adrenalin for its receptor may be described as the concentration required to produce a half-maximal response. This is approximately $10^{-6}$ mol l$^{-1}$. In pharmacology this value is more often expressed as $pD_2$, which is the negative logarithm of the molar concentration required to give a half-maximal response. Hence in this case the $pD_2$ value for adrenalin is 6 (Figure 3.30).

▷ If a ligand has a high $pD_2$ value, is its affinity for the receptor high or low?

▶ High. The lower the concentration required to produce a half-maximal response, the greater the $pD_2$.

The effect of the addition of propanolol on the adrenalin dose–response curve is also shown.

▷ How does this drug effect the $pD_2$ value?

▶ The $pD_2$ decreases to 4, so a higher concentration ($\times 100$) of adrenalin is required to produce a half-maximal response. This is because propanolol behaves as an antagonist, blocking access of β-receptors to adrenalin, thus preventing their activation.

## Summary of Section 3.7

The cell membrane is a physical barrier for many external stimuli and is a major site for the process of signal transduction, which enables a cell to respond to its surrounding environment.

Specialized intracellular communication systems located in the cell membrane respond to a diverse range of chemical, mechanical, electrical and light stimuli. Lipophilic substances, which are able to penetrate the cell membrane, interact with intracellular receptors that influence gene expression in the nucleus.

Following the detection of an external signal, receptors relay the information to a series of downstream components and this ultimately results in the biological response. The signal may amplified as it is transferred from one component to the next via either conformational changes or covalent modification. Termination mechanisms ensure that the cell maintains its sensitivity to the changing environment.

Ion channels may be gated by chemical, electrical or stretch stimulation, altering membrane permeability and initiating rapid cellular responses. The effects mediated by steroid receptors take longer to develop because the responses require transcription and translation. Some receptors are coupled to intermediary transducer molecules called G proteins. G protein activation exerts conformational changes in ion channels or enzymes present in the cell membrane. These enzymes catalyse the formation of second messengers such as cAMP (discussed in Section 3.9).

Another class of receptors, characterized by only one membrane-spanning region, possess intrinsic effector activity in the form of an enzymic domain which is stimulated on ligand binding. Tyrosine kinase receptors are the most common form of this type of receptor and are activated by insulin or growth factor polypeptides. Intracellular signalling is achieved through a series of phosphorylation cascades, which mediate the covalent modification of many different intracellular proteins such as enzymes, transporters and regulators of gene expression. Phosphorylation reactions are catalysed by kinases which act on tyrosine, serine or threonine residues, while phosphatases remove phosphate groups and so reverse these effects.

Some chemical ligands interact with more than one receptor protein, although the binding is highly stereospecific. Compounds that are structurally analogous to ligands may also bind specifically to receptor proteins, interacting in an agonistic or antagonistic fashion. Such effects form the basis of action of many drugs, and are a major area of pharmacological and pharmaceutical research. Receptors have their own particular signalling pathways, which can interact either positively or negatively to produce finely-tuned biological responses in the face of multiple external stimuli.

## 3.8   The role of G proteins in signal transduction

The discovery that all cells contain membrane-bound proteins that are intimately involved in the process of signal transduction (although are not receptors), was sparked by a rather unexpected experimental finding.

In the mid-1970s, Alfred Gilman and Elliott Ross, at the University of Virginia, were investigating the connection between adrenalin receptors, the membrane-bound enzyme adenylyl cyclase (which was known to generate the second messenger, cAMP) and GTP (Figure 3.31), a nucleotide which had also been shown to be an obligatory requirement for effective signalling by adrenalin across the cell membrane.

One particular study had focused on an unusual type of cell, called a $cyc^-$ mutant, which was capable of binding adrenalin, but appeared to lack any functional adenylyl cyclase. Hence, the addition of adrenalin to an extract of membrane proteins isolated from these $cyc^-$ cells failed to stimulate cAMP synthesis (Figure 3.32a), although this could be achieved in membranes which had been isolated from the wild-type ('normal') counterparts.

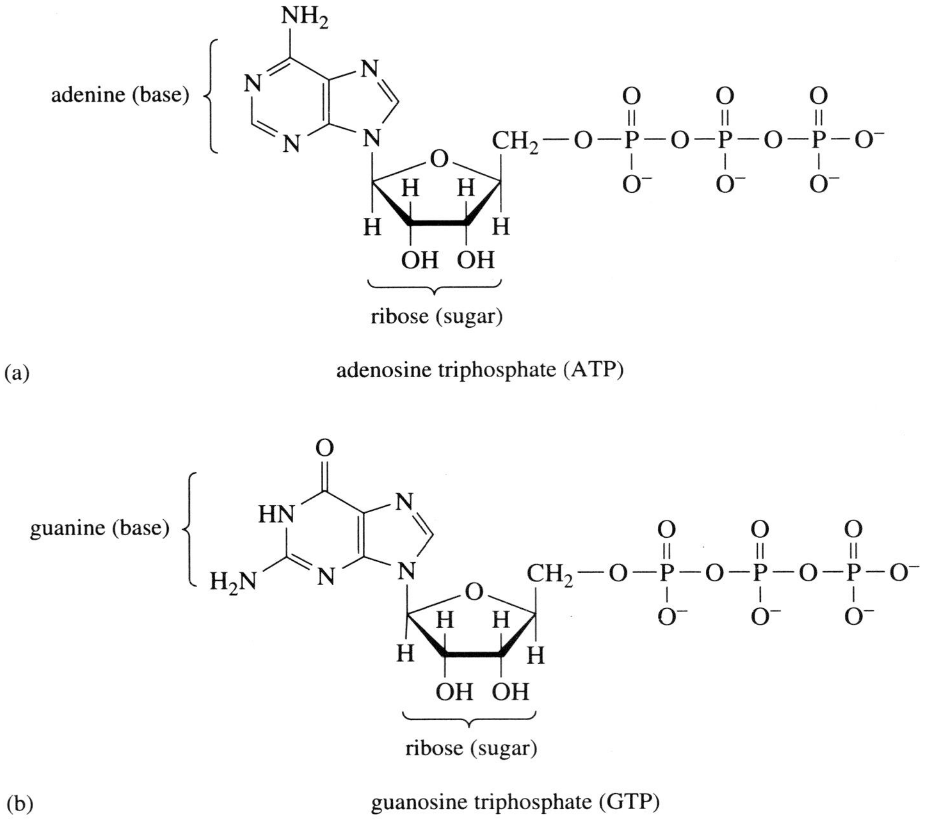

**Figure 3.31** A comparison of the formulae of ATP and GTP.

(a) adenosine triphosphate (ATP)

(b) guanosine triphosphate (GTP)

An attempt was made to reconstitute a functional form of the adenylyl cyclase enzyme into membrane extracts from the *cyc⁻* mutants by adding the cell membrane fraction from wild-type cells which were known to synthesize cAMP (Figure 3.32b). It was found that this addition produced a working coupled system in the membrane mixture so that adrenalin was now able to increase cAMP levels. However, as a control group, a portion of the membrane extract prepared from wild-type cells was chemically modified in order to inactivate the adenylyl cyclase present in these cells (Figure 3.32c). The results of this experiment are summarized in Figure 3.32.

**Figure 3.32** Reconstitution studies on *cyc⁻* mutants. (a) Membrane extracts from *cyc⁻* cells. (b) Membrane extracts from cyc⁻ cells supplemented with wild-type cell extracts. (c) Membrane extracts from *cyc⁻* cells supplemented with wild-type extracts in which adenylyl cyclase had been inactivated.

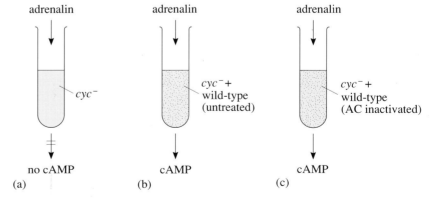

▷ Do *cyc⁻* mutants lack functional adenylyl cyclase?

▶ No. The experiment in Figure 3.32c indicates that it is not the adenylyl cyclase from the wild-type membrane extract which is restoring the ability of the *cyc⁻* cells to synthesize cAMP in the presence of adrenalin.

This observation led the investigators to hypothesize that another membrane component must be involved in the coupling of adrenalin β-receptors to the production of a second messenger. Subsequent experimental studies have identified the 'missing link' as $G_s$, a multisubunit GTP-dependent protein which specifically stimulates adenylyl cyclase.

$G_s$ is just one of over 20 different types of G protein that have, so far, been isolated and characterized. They receive signals from more than a hundred different sorts of receptor (including many which respond to odour and taste) and relay messages to a collection of enzymes (that synthesize second messengers) or to ion channels (Table 3.5). In addition, there are 'orphan' G proteins: the genes coding for these have been identified by cloning techniques, but the receptors and the downstream components with which they interact have yet to be established. The G protein-controlled regulatory network which operates across the cell membrane is often compared to a complex switchboard. There is a opportunity for either divergence or convergence of interactions at both the receptor–G protein level and at the G protein–second messenger level (Figure 3.33), and this endows cells with the potential for an extraordinary repertoire of responses. In addition, cells can also coordinate and finely-tune such responses to a range of environmental stimuli.

**Figure 3.33** The G protein switchboard. (a) Convergence of incoming signals. The incoming signals from ligands A and B coincide either at the same G protein (GPr) (i) or at the same target enzyme (ii). (b) Divergence of an incoming signal, i.e. the activation/inhibition of more than one second messenger system: (i) by multiple G protein activation of effectors; (ii) by multiple effector activation/inhibition. (*Note*: the G protein(s) could also interact with a target ion channel.)

(a) convergence of incoming signals

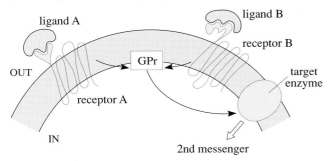

(i) on G protein

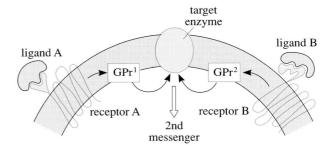

(ii) on target enzyme

(b) divergence of incoming signal

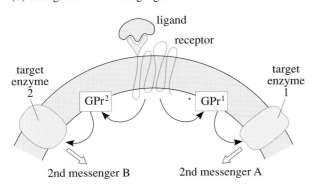

(i) by multiple G protein activation of effectors

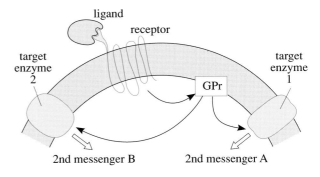

(ii) by multiple effector activation/inhibition

All G proteins have the same basic structural organization: they consist of three distinct subunits (α, which possesses the guanine nucleotide binding site, β and γ), and are free to move within the fluid phase of the phospholipid bilayer, either as the αβγ trimeric complex (when binding GDP) or as two segregated components, α (when binding GTP) and βγ (Figure 3.34). The βγ complex, which never dissociates, exhibits a particularly high affinity for the phospholipid bilayer due to the presence of a long hydrocarbon chain covalently attached to the γ subunit.

▷ How can this help to anchor the complex in the phospholipid bilayer?

▶ The hydrocarbon chain favourably interacts with the hydrophobic core of the membrane because it too is hydrophobic.

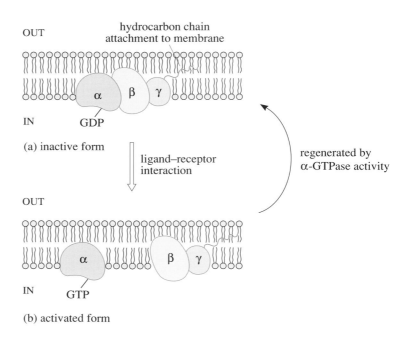

**Figure 3.34** The structure of G proteins. (a) The inactive trimeric form (GDPαβγ). (b) The activated form, which splits into two components, αGTP and βγ. Regeneration of the inactive trimer occurs via GTP hydrolysis (α-GTPase activity). Note that G proteins are located on the cytosolic side of the membrane.

In order to transmit information across the cell membrane, G proteins, like receptors, must be activated. In turn, this process of G protein activation either stimulates or inhibits an enzyme-linked second messenger system or an ion channel – these are the amplifiers shown in Figure 3.18 – and this ultimately generates a biological response.

In their inactive ('off') state, G proteins are found in the αβγ trimeric form, with GDP bound to the α subunit (Figure 3.35a). When a receptor is activated by either light or a chemical stimulus, an interaction with a specific G protein occurs because of the enhanced affinity of the receptor for the αβγ complex. As a consequence of receptor binding, GDP is ousted from the α subunit and replaced by GTP (Figure 3.35b). Nucleotide exchange is followed by G protein dissociation to the αGTP and βγ forms, and also by a decrease in their affinity for the activated receptor so that the two components regain their mobility within the membrane.

The association of αGTP with a specific amplifier forms the basis of the second leg of the relay (Figure 3.35c). The target of the αGTP subunit varies depending on the type of G protein that is activated (Table 3.5). In the case of adenylyl cyclase, this enzyme may be either activated by the α subunit of $G_s$ or inhibited by the α subunit of $G_i$: hence those receptors that activate $G_s$ increase the cAMP concentration inside cells, while those that activate $G_i$ cause cAMP levels to decrease. In some situations, βγ performs a transducing role, either by direct interaction with specific second messenger systems or indirectly, by interfering with free α subunits and thus preventing their downstream effects.

The regeneration of the inactive GDPαβγ trimer is an automatic event. The α subunit is also an enzyme: it possesses *intrinsic GTPase* activity. After a few seconds, the terminal phosphate on the bound GTP is hydrolysed to GDP, which remains associated with the α subunit (Figure 3.35d). The G protein thereby switches itself off, since the α subunit is no longer capable of interacting with effectors, and instead reassociates with the βγ dimer to form the original GDPαβγ complex (3.35a). The trimer is thus available for subsequent reactivation by the same or another activated receptor, thus providing scope for an amplification of the original stimulus.

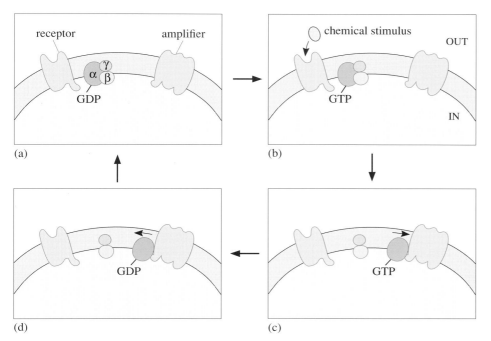

**Figure 3.35** How G proteins are switched on and off. (a) In their resting state, G proteins are bound to GDP and have no contact with receptors. (b) When a stimulus (light or chemical) activates a receptor, the receptor causes the G protein to exchange GDP for GTP, which activates the G protein. (c) The G protein then dissociates, after which the GTP-bound α subunit diffuses along the membrane and binds to an amplifier, thereby activating it. (d) After a few seconds, the α subunit converts GTP to GDP, thereby inactivating itself. The α subunit will then reassociate with the βγ complex (a).

You should now appreciate the critical role that the cycle of GDP/GTP exchange plays in determining the magnitude of a G protein-linked biological response. Many experiments have exemplified its importance, including some that have studied the effects of non-hydrolysable GTP analogues which are resistant to the effects of GTPases (Figure 3.36).

▷ What will happen to the rate of cAMP synthesis if the analogue p(NH)ppG is added to the membrane extracts shown in Figure 3.32a and b?

▶ No cAMP will be synthesized by the *cyc⁻* cell membranes (Figure 3.32a), since they lack $G_s$. When $G_s$ is present (Figure 3.32b), p(NH)ppG displaces GDP from its binding site to the α subunit (due to β-receptor activation by adrenalin). The $G_s$ αp(NH)ppG complex then permanently activates adenylyl cyclase, since the GTPase activity has no effect in this case. Thus maximal rates of cAMP synthesis would be observed.

### 3.8.1  $G_i$ and $G_s$ response to toxins

In the presence of bacterial toxins, the α subunits of some classes of G protein undergo covalent modification by the addition of an ADP-ribose group (usually donated from $NAD^+$) to a specific amino acid residue on the α polypeptide chain (Figure 3.37). The effect of toxin-induced covalent modification on G protein function is detrimental, but not all types of G protein (shown in Table 3.5) are equally susceptible.

GTP

GTPγS

p(NH)ppG

**Figure 3.36** The structure of GTP and of two non-hydrolysable GTP analogues.

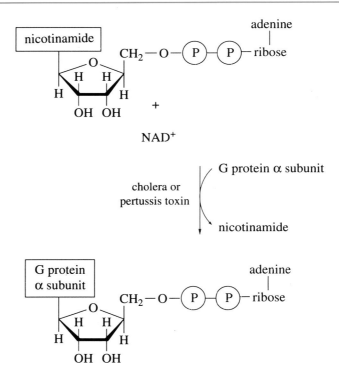

**Figure 3.37** Covalent modification of G protein ($G_s$ or $G_i$) α subunits by the addition of an ADP-ribose group from $NAD^+$.

▷ Cholera toxin catalyses the addition of ADP-ribose to a specific arginine residue on the α subunit of $G_s$, causing an inhibition of its intrinsic GTPase activity. How would this affect $G_s$ function?

▶ $G_s$ would be permanently switched on, as the deactivation mechanism would be disabled.

As a consequence, any cholera toxin which enters intestinal cells triggers the accumulation of excess intracellular cAMP. This causes the cells to secrete large amounts of electrolytes and water into the lumen of the gut; the diarrhoea which ensues may lead to lethal dehydration.

In contrast, pertussis toxin (released from the bacterium *Bordetella pertussis* which causes whooping cough) catalyses the covalent attachment of ADP-ribose to a cysteine residue located near the C-terminus of the $G_i$ α subunit. The modified $G_i$ α subunit is no longer able to interact with a receptor, so the overall effect of this toxin is to uncouple the signal transduction process.

The involvement of G proteins in transmembrane signalling can be investigated using these bacterial toxins. For example, a response that involves $G_s$ would be considerably enhanced in the presence of cholera toxin, and a response that operates through $G_i$ would be inhibited in the presence of pertussis toxin. Other types of G protein are affected by neither or both of the toxins.

### 3.8.2 Some examples of G protein-linked systems

#### The response to light

The signal transduction system that mediates the detection of light by the photosensitive cells in the retina of the mammalian eye is well characterized, and is illustrated in Figure 3.38. Light-activated rhodopsin is linked to its specific intracellular effector via a G protein called *transducin* ($G_t$). In turn, the activity of the enzyme cGMP phosphodiesterase is increased, causing the cGMP levels in the cell to decrease. Because cGMP keeps Na$^+$ channels open (recall from Table 3.4 that these channels are ligand-gated), its hydrolysis to GMP results in a closure of the channels. Na$^+$ can therefore no longer enter the cells so that the negative charge inside increases, eventually leading to hyperpolarization. This signal is eventually relayed from the retina via the optic nerve to the brain.

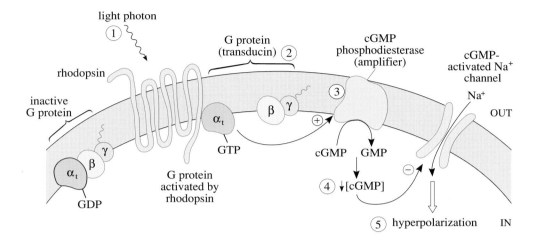

**Figure 3.38** The rhodopsin–G protein system. (1) Light falls on a molecule of rhodopsin, inducing a conformational change leading to receptor activation. (2) Rhodopsin then interacts with G protein (transducin, G$_t$) promoting GDP exchange for GTP and the formation of the two components, $\alpha_t$GTP and $\beta\gamma$. (3) $\alpha_t$GTP interacts with cGMP phosphodiesterase (amplifier), causing activation: the phosphodiesterase hydrolyses cGMP to GMP so the cGMP level in the cell decreases. (4) cGMP maintains Na$^+$ channels open (transmitter-gated by intracellular ligand), so the decrease in cGMP causes the Na$^+$ channels to close. (5) Channel inhibition results in hyperpolarization, leading to the generation of an electrical signal to the brain. (In this and subsequent figures a plus sign denotes activation and a minus sign inhibition.)

## The regulation of contractility in the heart

The contractility of heart muscle is mediated by the intracellular concentration of cAMP. You have certainly experienced the effects of adrenalin on heart beat (the hormone is produced in the body as part of the 'flight or fight' reaction to stress or danger): the increased force and rate of contraction is due to the activation of $\beta$-receptors, which act via G$_s$ and stimulate an increase in cAMP levels. The neurotransmitter acetylcholine, acting through muscarinic receptors, activates G$_i$ and opposes the effects of adrenalin in two ways: first, it lowers the cAMP concentration in heart cells by inhibiting adenylyl cyclase and second, it increases the permeability of heart cells to K$^+$ ions, which flow out of the cell and impede contraction. The activated G$_i$ protein splits into $\alpha_i$GTP and $\beta\gamma$ and then 'homes in' on two distinct amplifier systems: adenylyl cyclase and a K$^+$ channel (Figure 3.39). The $\alpha_i$GTP target is the channel, which opens in response to the interaction. However, in this system, $\beta\gamma$ is also considered to play an important role in the transduction process by associating with any $\alpha$ subunits that have been released from G$_s$. In this way, the activation of adenylyl cyclase by an interaction with $\alpha_s$GTP is prevented. Hence, although $\alpha$GTP is usually considered to be the business end of a G protein molecule, the $\beta\gamma$ subunits also contribute to the signalling process.

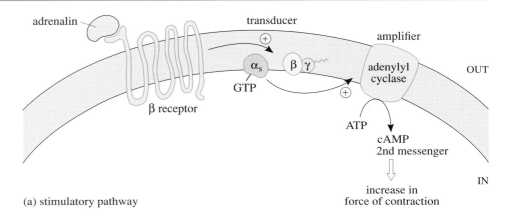

(a) stimulatory pathway

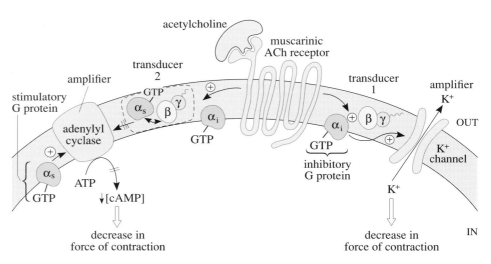

(b) inhibitory pathways

*Figure 3.39* The role of G proteins in the regulation of contractility in the heart. (a) Stimulatory pathway. The activated β-receptor stimulates G protein dissociation; the $\alpha_s$GTP subunit activates adenylyl cyclase, elevating the cytosolic concentration of cAMP. This increases the force of contraction as a result of downstream events. (b) Inhibitory pathways. Acetylcholine (via muscarinic receptors) decreases the force of contraction in two ways. First, it activates $G_i$, which opens $K^+$ channels through an interaction of the channel with $\alpha_i$GTP (transducer 1). Second, the same G protein contributes to the inhibition of adenylyl cyclase. In this case, the βγ released by G protein activation 'mops up' any stimulatory $G_s$ α subunits, attenuating the activation of adenylyl cyclase and thereby quietening the stimulatory pathway.

### 3.8.3 Signal amplification and termination

What are the advantages of the G protein system of signalling? The presence of intermediary G proteins permits **signal amplification**. G proteins magnify the effect of receptor activation in two ways (Figure 3.40). First, one activated receptor may activate (and sometimes reactivate) many G protein molecules in quick succession before ligand dissociation ensues. In the case of rhodopsin, one light-activated receptor stimulates more than 500 molecules of transducin ($G_t$) almost simultaneously and this accounts for the very high sensitivity of the photoresponsive cells in the eye, which can detect as little as a single photon of light. Second, when one molecule of αGTP interacts with an enzyme such as

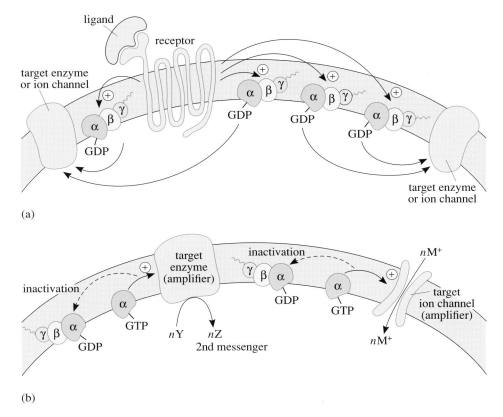

(a)

(b)

***Figure 3.40*** Amplification by G proteins. (a) One activated receptor can stimulate GDP/GTP exchange in many G proteins molecules in the cell membrane. (b) One activated G protein can stimulate/inhibit a target enzyme and regulate the number of second messenger molecules (Z) produced before deactivation. Similarly, one activated G protein can allow many ions ( $M^+$) to enter the cell before deactivation.

adenylyl cyclase, many molecules of cAMP are synthesized in the few seconds before auto-inactivation (via the $\alpha$ subunit's intrinsic GTPase activity) to the GDP form sets in. Similarly, an ion channel allows the entry or exit of many ions before 'switch-off' occurs.

G proteins also exert control on **signal termination** and as a consequence of this, the magnitude of the biological response of a cell to an external stimulus (i.e. its sensitivity) can be modulated.

$\triangleright$ What would happen to a G protein-linked response if the time taken for GTPase activation increased?

$\blacktriangleright$ The response to a given stimulus would increase because there is more time available for the $\alpha$GTP to interact with the downstream amplifier.

The association of the G protein with an activated receptor may be controlled by the phosphorylation of serine and threonine residues in the receptor and/or G protein $\alpha$ subunit (Figure 3.41). The ATP-linked phosphorylation of the OH groups on the side chains of the serine and threonine is associated with a reduction in the efficacy of coupling due to a decrease in the number of receptors at the cell surface (see Section 3.6.1) and/or a decrease in the affinity of the G protein for the activated receptor (termed **desensitization**). The decrease in receptor number and/or G protein affinity is induced in the presence of very high concentrations of ligand, and results in a decrease in the cell's response to the stimulus (sometimes referred to as *down-regulation*).

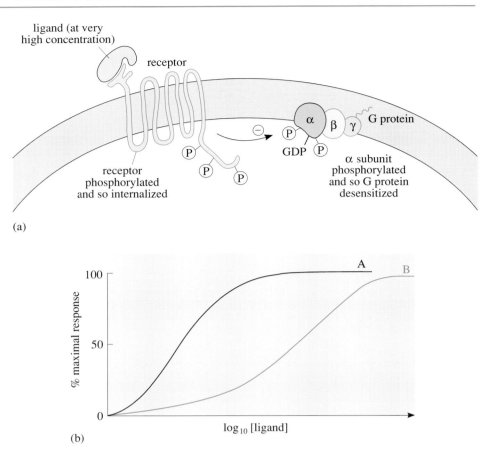

**Figure 3.41** (a) Mechanisms of desensitization in G protein-linked systems. In the presence of very high concentrations of ligand, phosphorylation of receptor and G protein occurs. Phosphorylation of the receptor promotes its internalization, decreasing the receptor population in the cell membrane. Phosphorylation of the G protein α subunit decreases the affinity of the αβγ complex for the receptor, i.e. the G protein becomes desensitized. (b) The effect of densitization on biological responses. Curve A: normal dose–response to a ligand. Curve B: dose–response after cells have been exposed to a very high concentration of ligand – the cells are less sensitive to ligand, but note that a maximal response can be achieved when very high concentrations of ligand are given.

### 3.8.4  The GDP/GTP exchange protein superfamily

All the α subunits of G proteins are members of a superfamily of intracellular proteins, united by the fact that they possess a guanine nucleotide binding site and that they are regulated by GTP/GDP exchange. Other members of this class of proteins include *ras proteins,* which are also membrane-bound and which are involved in the control of cell division, and the cytoplasmic *initiation factors,* which control the efficiency of ribosomal protein synthesis. The broad spectrum of biological functions carried out by GTP/GDP exchange proteins suggests that they all may have originated from a single ancestral gene, which duplicated and then diversified.

### 3.8.5  The clinical potential of G protein research

G proteins control the specialized functions of most cells (Table 3.5). For this reason, they are now the focus of much research, amidst speculation that they may contribute to pathological conditions as diverse as heart failure, diabetes and depression. If this proves to be the case, the scope for therapeutic intervention is certain to increase, since the mode of action of G proteins is well established and knowledge on the subject continues to expand.

### Summary of Section 3.8

Many receptors employ G proteins as a means of signal transduction across the cell membrane; they are usually characterized by the presence of seven membrane-spanning domains. The varied G proteins located in the phospholipid bilayer of the cell membrane have a common structural organization: an α subunit, which binds GDP or GTP, along with one β and one γ subunit. In the 'resting' state, G proteins are found as trimers, complexed with GDP. Displacement of GDP by GTP is promoted by receptor activation and leads to G protein dissociation and the interaction of the αGTP (or in some cases βγ) with specific target amplifiers such as enzymes which synthesize second messengers, or with ion channels. Termination is ensured by an auto-inactivation mechanism which involves the α subunit's intrinsic GTPase activity.

G protein signal transduction is associated with amplification at each stage. The coupling process may be physiologically modulated by the phosphorylation of receptors and G proteins; bacterial toxins and non-hydrolysable GTP analogues are used experimentally to investigate the role of G proteins in cellular systems. Well-studied examples of G protein-linked systems include the stimulation of cAMP synthesis by adrenalin (via β-receptors) and the response of the photoreceptors in the eye to light.

## 3.9 Some important second messenger systems

### 3.9.1 Adenylyl cyclase and cAMP

Adenylyl cyclase, the enzyme that synthesizes the second messenger cAMP from ATP, is an important amplifier located in the cell membrane of all eukaryotic cells. It is a large protein with 12 transmembrane domains. Table 3.6 highlights some of the chemical stimuli which, in mammalian systems, employ adenylyl cyclase as a means of intracellular communication.

A number of different forms of adenylyl cyclase have now been identified. Two hydrophilic cytosolic domains of the enzyme contain the active site. The $K_M$ values of most eukaryotic adenylyl cyclases are between 50 and 100 µmol l$^{-1}$, while the intracellular concentration of ATP is approximately 100 times higher, at 5–10 mmol l$^{-1}$ (Figure 3.42).

**Table 3.6** Examples of stimuli that regulate the activity of adenylyl cyclase.

| Stimulatory | Inhibitory |
| --- | --- |
| adrenalin (via β-receptors) | acetylcholine (via muscarinic receptors) |
| antidiuretic hormone (ADH) (also called vasopressin) | insulin |
| glucagon | |
| luteinizing hormone (LH) | |
| odorants (detected via smell receptors) | |

(Further information on hormones is provided in Chapter 4.)

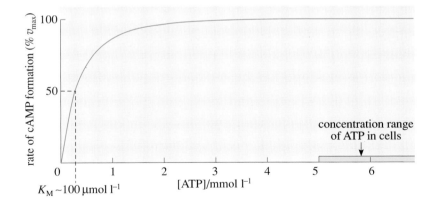

**Figure 3.42** The kinetics of the adenylyl cyclase reaction.

▷ What do the values of $K_M$ and intracellular ATP concentration tell you about the nature of the reaction catalysed by adenylyl cyclase? (Figure 3.42 should help you.)

► Adenylyl cyclase is always saturated with substrate.

Saturation is common among second messenger enzyme systems, since it ensures that the rate of formation of a second messenger such as cAMP is not limited by the concentration of substrate (in this case, ATP) and is instead determined solely by the activity of the enzyme.

▷ What signalling component (that you have already encountered) regulates the activity of adenylyl cyclase?

▶ G proteins: adenylyl cyclase is stimulated by $G_s$ and inhibited by $G_i$. G proteins are regulated by many different surface receptors and activated in response to ligand binding.

Historically, cAMP was the prototype **second messenger**. The 'second messenger' concept was developed by Earl Sutherland in the 1950s during his pioneering studies on transmembrane signalling. He was interested in how the hormones adrenalin and glucagon, both hydrophilic compounds, managed to communicate across the phospholipid barrier of liver cells and stimulate the breakdown of glycogen. A heat-stable soluble factor, later identified as cAMP, was isolated from an extract of liver cell membranes when incubated with hormone, and this in turn was shown to activate the enzymes involved in regulating glycogen breakdown. Sutherland postulated that through binding to receptors, the hormones stimulate the synthesis of the heat-stable chemical mediator, which then acts as an intracellular signal. Hence there is no need for hydrophilic ligands such as adrenalin and glucagon to penetrate the cell in order to exert their biological effects. (Note that in adipose tissue the same receptor stimulates the breakdown of stored fat (lipolysis).)

How exactly does cAMP affect cellular processes? Many of its biological actions stem from an initial interaction with another intracellular enzyme complex, the cAMP-dependent protein kinase, also known as protein kinase A (PKA).

PKA contains two types of subunit, one catalytic (C) and one regulatory (R). In the presence of very low concentrations of cAMP, PKA exists as the $R_2C_2$ tetramer and is inactive: the catalytic subunits are immobilized because they are bound (non-covalently) to the regulatory subunits (Figure 3.43). However, when the cAMP concentration is increased (say around 10-fold, from 1 to $10\ \mu mol\ l^{-1}$) it too binds to the regulatory subunits and this allosteric interaction causes conformational changes in the PKA that reduce the affinity of R for C. The catalytic subunits dissociate, and are then free to exert their biological effects.

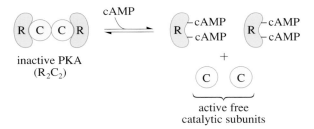

*Figure 3.43* The structure of cAMP-dependent protein kinase (PKA) and its regulation by cAMP.

As its name suggests, protein kinase A catalyses phosphorylation reactions (using ATP as a phosphate donor) within the cell. Its targets are the serine and threonine residues present in a diverse range of intracellular proteins. Many of these proteins are enzymes, which may be either activated or inhibited by phosphorylation. Other target proteins for the action of PKA include receptors (found embedded in the cell membrane or within the cytosol), G protein subunits, transporters and actin binding proteins (which regulate the microfilament structure of the cytoskeleton). You should now begin to appreciate how cAMP is able to cause such a spectrum of biological effects in different cells.

Figure 3.44a illustrates the sequence of events that lead to the cAMP-mediated stimulation of glycogen breakdown in skeletal muscle. The mobilization of glycogen to glucose 6-phosphate (Figure 3.44b) provides a source of energy for muscle contraction during exercise.

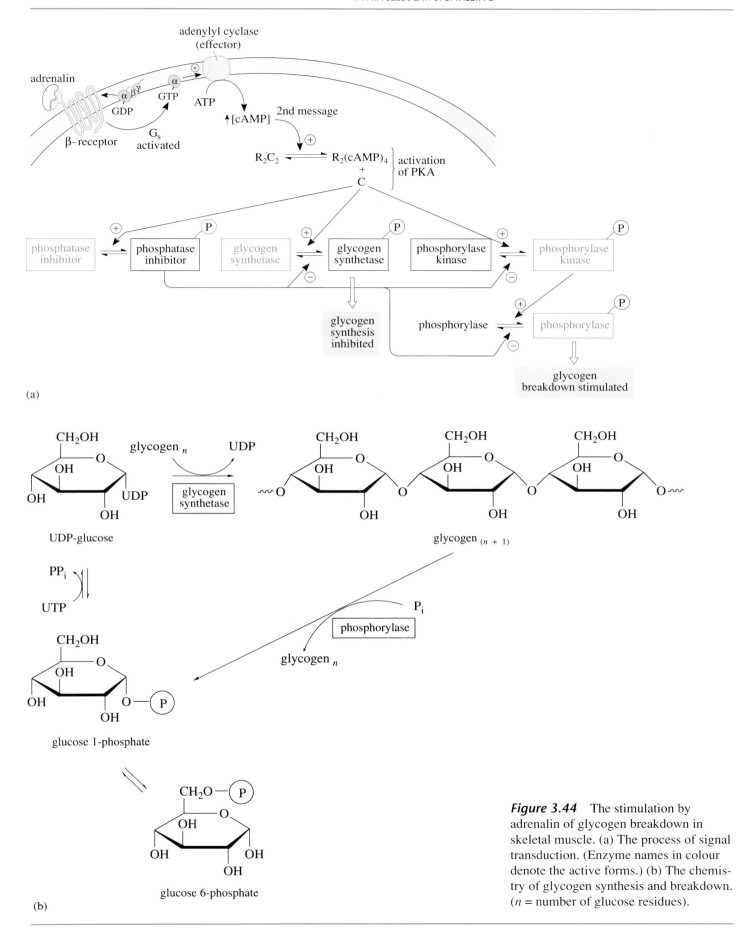

(a)

(b)

***Figure 3.44*** The stimulation by adrenalin of glycogen breakdown in skeletal muscle. (a) The process of signal transduction. (Enzyme names in colour denote the active forms.) (b) The chemistry of glycogen synthesis and breakdown. (*n* = number of glucose residues).

The occupation of β-receptors by adrenalin activates $G_s$, which causes a rise in intramuscular cAMP. The subsequent activation of PKA leads to the phosphorylation of three key intracellular enzymes (Figure 3.44a):

o *glycogen synthetase* is phosphorylated and inactivated

o *phosphorylase kinase* is phosphorylated and activated

o *phosphatase inhibitor* is phosphorylated and activated

The metabolism of glycogen is controlled by two enzymes (Figure 3.44b): one catalyses glycogen synthesis (glycogen synthetase) while the other acts in opposition and promotes glycogen breakdown (phosphorylase, so-called because the hydrolysis of glycogen involves the attachment of a phosphate group to each glucose molecule released).

▷ In order to achieve a significant rate of glycogen breakdown in muscle what effects on these two key enzymes must occur?

▶ Phosphorylase must be stimulated and glycogen synthetase must be inhibited.

The synthetic pathway is directly inhibited by the action of PKA. However, the stimulation of glycogen breakdown is indirect, and mediated by a phosphorylation cascade. The activation of phosphorylase kinase by PKA is the first step in this cascade.

▷ What is the effect of activating phosphorylase kinase?

▶ Phosphorylase is covalently modified by the addition of a phosphate group and the enzyme is activated, causing a stimulation of glycogen breakdown.

One additional effect of PKA in muscle is the activation of a phosphatase inhibitor. How does the inhibition of phosphatases contribute to the overall response? Intracellular phosphatases oppose the effect of cAMP on glycogen synthetase, phosphorylase kinase and phosphorylase (Figure 3.44a). By inhibiting their actions, a greater stimulation of glycogen mobilization is observed.

Not all the effects of cAMP are mediated through PKA. cAMP can also interact with ion channels directly. For example, in olfactory neurons, cAMP binds to and opens $Na^+$ channels (see Table 3.4) and this important action allows us to detect different odours.

What mechanisms exist for terminating the effects of cAMP? A decrease in the rate of cAMP synthesis by adenylyl cyclase may be mediated by a decrease in the activation of $G_s$ or an increase in the activation of $G_i$, or both. In addition, cAMP already formed may be hydrolysed to inactive AMP by the action of phosphodiesterases. One type is particularly important in 'switching off' the actions of cAMP: its $K_M$ (~ 1 μmol l$^{-1}$) is similar to the intracellular concentration range of cAMP so it is able to respond effectively to physiological changes in cAMP concentration (Figure 3.45).

Thus cAMP synthesis and breakdown both contribute to the intracellular cAMP concentration and serve as a means of controlling biological responses and their termination (Figure 3.46). Adenylyl cyclase and cAMP phosphodiesterase activity may also be manipulated pharmacologically. For example, forskolin (an organic compound isolated from the root of an Indian herb) is a very potent activator of adenylyl cyclase and acts independently of a G protein. In contrast, caffeine and theophylline bind to and inhibit phosphodiesterases.

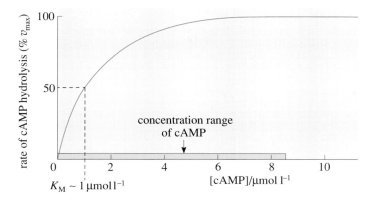

**Figure 3.45** The kinetics of cAMP phosphodiesterase activity.

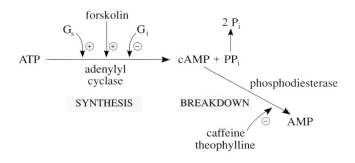

**Figure 3.46** The synthesis and breakdown of cAMP showing the effects of some activators and inhibitors.

The addition of forskolin to isolated liver cells stimulates the mobilization of glycogen. This is because cAMP (the concentration of which is elevated, due of the direct effect of forskolin on adenylyl cyclase) stimulates phosphorylase and inhibits glycogen synthetase by the same pathways that operate in skeletal muscle (Figure 3.44a). Under normal conditions, the cAMP-mediated breakdown of liver glycogen is stimulated by the hormone **glucagon**. In the liver, the glucose 6-phosphate liberated is converted into glucose (rather than used to generate energy, as in muscle). The glucose is then released into the bloodstream to maintain a basal concentration during conditions of fasting.

▷ In an experiment, isolated liver cells were incubated with glucagon for 30 minutes and the rate of glucose release measured. How would the rate of glucose release be affected if the experiment was repeated with cells that had been preincubated with caffeine?

► The rate of glucose release would be greater because more cAMP would be synthesized in response to glucagon if phosphodiesterase activity was inhibited.

The inhibition of phosphodiesterase activity potentiates the biological effects of any ligand that employs cAMP as a signal transduction system and is therefore a useful probe for investigating if a receptor is coupled to adenylyl cyclase.

The G protein-linked receptor activation of adenylyl cyclase and subsequent PKA cascades or opening of ion channels are very rapid events within cells. Hence most of the biological effects that involve cAMP are immediate, and observed within seconds or minutes of the stimulus. However, recent studies suggest that the catalytic subunit of PKA may also mediate long-term effects within cells by regulating gene expression. A specific protein (the cAMP response element) has been located on nuclear DNA and this may control the rate of transcription of a specific gene by the binding to the catalytic subunit of PKA. This strategy of intracellular communication is commonly exploited by lipophilic hormones (e.g. steroids), which act via intracellular receptors. All nuclear-based effects are characterized by latent responses to stimuli because a longer time period (minutes/hours/days) is required for the changes in the rate of synthesis of specific proteins within the cell to occur. They are often associated with cell development and differentiation. Interestingly, cAMP also affects gene expression in prokaryotes. In bacteria, it functions as a 'hunger signal' so that cAMP levels increase when glucose is scarce. The cAMP then interacts with DNA binding proteins and switches on genes coding for enzymes that catabolize alternative substrates.

### 3.9.2  Guanylyl cyclase and cGMP

You have already encountered the membrane-bound form of guanylyl cyclase. It is the natriuretic peptide hormone receptor, a membrane-spanning protein that includes a cytosolic guanylyl cyclase domain (Figure 3.25). This receptor was the first protein shown to directly catalyse the formation of a chemical second messenger (i.e. cGMP) in response to extracellular ligand binding (so is analogous to the transmitter-gated ion channel mode of signal transduction).

In addition to its role in natriuretic receptors, guanylyl cyclase is also found free in the cytosol. This so-called 'soluble' form of guanylyl cyclase is activated by the lipophilic mediator nitric oxide (NO) whose multiple roles began to be discovered only in the early 1990s, and which was designated 'molecule of the year' in 1992 by the journal *Science*. You will learn more of the functions of this molecule in Chapter 4.

cGMP has two key modes of action within cells:

*1   It is a regulator of ion channels*    The regulation of ion channel activity by cGMP is, as you have already learned, particularly physiologically important in the photosensitive cells of the eye.

▷    What is the major mechanism for controlling the intracellular concentration of cGMP in these cells and what is its effect?

▶    The activity of cGMP phosphodiesterase regulates cGMP levels (although guanylyl cyclase is, of course, essential for cGMP synthesis). Light activates the phosphodiesterase via a specific G protein (transducin, $G_t$), thus causing a decrease in the intracellular cGMP concentration. Since cGMP is necessary to keep $Na^+$ channels open, the detection of light photons at the retina is associated with the closure of the channels, which then leads to membrane hyperpolarization.

*2   It activates a specific cGMP protein kinase (PKG)*    This mode of action of cGMP is akin to that of cAMP (via PKA). PKG also targets serine and threonine residues of intracellular proteins, but it has a more restricted repertoire of substrates than PKA and seems to only exert significant physiological effects in few cell types, namely smooth muscle, blood platelets and certain areas of the brain. PKG has a different structural organization to PKA: it is a dimer composed of two identical subunits and has no accessory regulatory protein. Its catalytic action is thus stimulated by direct binding of cGMP.

In smooth muscle, an increase in intracellular cGMP facilitates relaxation. (The effect is promoted by NO.) The process of relaxation appears to require phsphorylation and hence activation of key target proteins, including $Ca^{2+}$ ATPase and a $Ca^{2+}$ regulatory protein found in the internal membranes of muscle cells, the sarcoplasmic reticulum (Figure 3.10). The resultant extrusion of $Ca^{2+}$ ions, either across the cell membrane or into intracellular stores, is considered to cause, at least in part, the lowering of cytosolic $Ca^{2+}$ concentration which is necessary to induce the relaxation of smooth muscle.

### 3.9.3 Phospholipase C, inositol trisphosphate and diacylglycerol

*Phospholipase C* (*PLC*, also known as phosphoinositidase C) is the enzyme responsible for generating two important second messengers, **inositol trisphosphate** ($\text{IP}_3$) and **diacylglycerol** (**DAG**). These two chemical mediators operate via distinct intracellular signalling systems which, as you will discover later, often interact to produce the observed biological response.

Phospholipase C may be activated by two classes of receptors (Table 3.7 *overleaf*). One type operate via a specific G protein (called $G_q$) and this in turn interacts with a membrane-bound form of the enzyme, causing stimulation (Figure 3.47a). Receptors that possess intrinsic tyrosine kinase activity are also able to increase the activity of phospholipase C, although a different form of the enzyme is stimulated in response to receptor binding. The autophosphorylation of specific tyrosine residues on these receptors provide a docking site for the binding and activation of a number of cytosolic proteins, including PLC (Figure 3.47b). Some receptors lacking an intracellular tyrosine kinase domain are still able to stimulate PLC in this way since they recruit other intracellular tyrosine kinases (so-called non-receptor tyrosine kinases) which act like a bridge between the receptor and the enzyme (Figure 3.47c). This phenomenon has been observed in a class of immune cells called T lymphocytes following the recognition of a foreign protein (the antigen) by a specific T cell receptor present on the cell surface.

The idea that extracellular ligands increase membrane phospholipid metabolism and, furthermore, employ it as a means of intracellular communication evolved from studies conducted by Mabel and Lowell Hokin in Sheffield and Montreal in

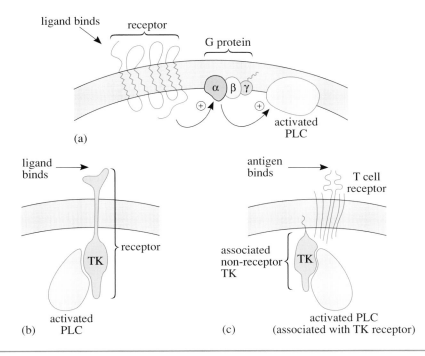

**Figure 3.47** The activation of phospholipase C: (a) G protein linked; (b) via a tyrosine kinase (TK) receptor – direct association; (c) indirect association via a tyrosine kinase 'link' protein (non-receptor TK).

*Table 3.7* Ligands and receptors that activate phospholipase C.

| Signal transduction through G protein-linked receptors | Signal transduction through tyrosine kinase-linked receptors | |
| --- | --- | --- |
| | Direct | Indirect* |
| noradrenalin/α-receptors | platelet-derived growth factor (PDGF) | antigen receptors in immune cells |
| acetylcholine/muscarinic receptors | epidermal growth factor (EGF) | |
| 5-hydroxytryptamine (5-HT, serotonin) | | |
| angiotensin II | | |
| antidiuretic hormone (ADH, vasopressin) | | |
| parathyroid hormone | | |
| gonadotrophin releasing hormone | | |

* via non-receptor tyrosine kinase docking proteins.

the 1950s. Their early work on slices of pancreas showed that acetylcholine stimulated the incorporation of radioactively labelled phosphate into phosphatidyl inositol, long before the identity of phospholipase C was known.

However, the events observed by the Hokins represent a resynthesis pathway, which is always stimulated following the receptor-coupled activation of phospholipase C. The enzyme actually hydrolyses a membrane phospholipid, and the substrate turns out to be a derivative of phosphatidyl inositol, *phosphatidyl inositol 4,5-bisphosphate ($PIP_2$)* (Figure 3.48). Phosphatidyl inositol (PI) itself is only a minor component of the cell membrane; it is more abundant within the inner leaflet of the bilayer but makes up only about 5% of the total phospholipid. $PIP_2$ is generated within the membrane by the action of specific kinases, but constitutes only a few per cent of the small PI pool, so as a precursor it is very scarce. An efficient recycling system maintains the $PIP_2$ concentration within the membrane (Figure 3.48). However, levels can become depleted as a result of prolonged high-dose stimulation and this results in an impairment of the transduction system. Thus, the magnitude of an $IP_3$-mediated response decreases; this is referred to as desensitization. (This type of desensitization is analogous to that discussed in Sections 3.6.1 and 3.8.2, and results in a dose–response curve similar to that shown in Figure 3.41b)

The reaction catalysed by phospholipase C is shown in Figure 3.49. Note the bond that is broken during hydrolysis: this is the target for all phospholipase C enzymes (although, as you will see, not all are specific for $PIP_2$).

▷ Examine the structures of $IP_3$ and DAG. Where in the cell do you think you might find these second messengers?

▶ $IP_3$ is a hydrophilic, negatively charged molecule which would favour the watery environment of the cytosol. In contrast, DAG is largely non-polar and so is retained within the bilayer (where it is formed).

So you can see that the so-called bifurcating (i.e. two-limbed) second messenger system generated by the action of phospholipase C is localized to two distinct

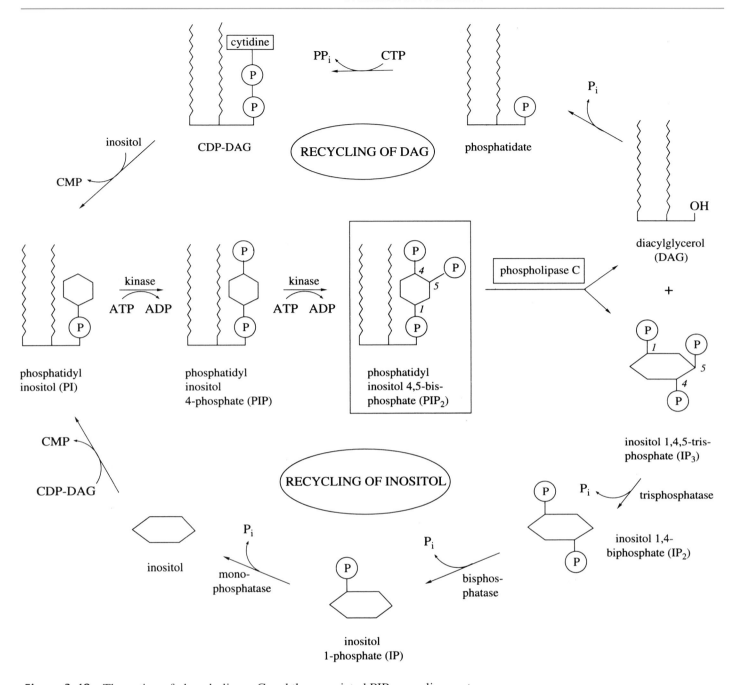

**Figure 3.48** The action of phospholipase C and the associated PIP$_2$ recycling systems.

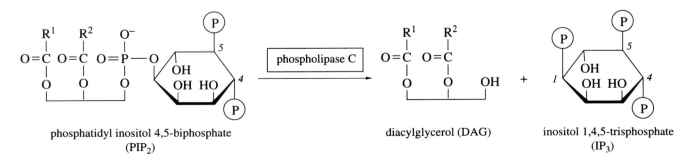

**Figure 3.49** The chemistry of PIP$_2$ hydrolysis. (R$_1$ and R$_2$ are the long, lipophilic fatty acid tails.)

regions of the cell: the lipid and aqueous phases. We shall now examine in more detail the precise roles that these second messengers play in the process of intracellular communication.

### Inositol trisphosphate and calcium

$IP_3$ controls a diverse range of cellular processes through its effects on intracellular calcium levels. For this reason, we shall first focus on the distribution of calcium ions inside a typical cell. The concentration of free calcium ions in the cytosol is very low, usually at a submicromolar concentration (of order $10^{-7}$ mol $l^{-1}$). In contrast, the extracellular level of calcium is about 1 000 times higher (Table 3.1).

▷ What maintains this concentration gradient?

▶ $Ca^{2+}$ ATPase located in the cell membrane.

Despite its low cytosolic concentration of calcium, the cell has considerable reserves which are found in two key organelles, the endoplasmic reticulum and the mitochondrion, as you may recall from Book 1, Chapter 1. These two organelles accumulate calcium by active transport processes. In the case of the endoplasmic reticulum, calcium storage is promoted by the presence of specific binding proteins such as *calsequestrin*. Hence the cell has two sources of calcium at its disposal, internal and external, which could, at least in theory, be employed for signalling purposes.

The effect of $IP_3$ on intracellular calcium levels may be investigated using fluorescent dyes. Cells incubated with these fluorescent compounds emit light at an intensity which is proportional to the cytosolic concentration of calcium. Figure 3.50 shows the effect of $IP_3$ on the fluorescence of cells preloaded with such a dye. (See also Plate 1.1.)

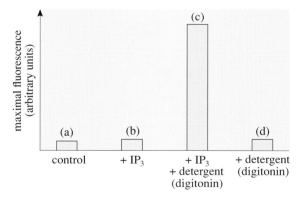

***Figure 3.50*** The effect of $IP_3$ on intracellular calcium levels. (a)–(d) show the effects of incubating isolated cells with the fluorescent dye Fura 2 under various conditions.

▷ Why do you think that $IP_3$ had no effect in experiment (b)?

▶ $IP_3$ was not able to enter the cell.

The data shown in Figure 3.50, experiment (c) illustrate that in the presence of a detergent which allows access to the inside of the cell, $IP_3$ causes an increase in the intracellular $Ca^{2+}$ concentration. What is the source of this rise in intracellular $Ca^{2+}$ concentration? Clearly, there are a number of possibilities. A major role for external calcium in the $IP_3$ response has been ruled out by the observation that in the presence of EGTA (ethylene glycol tetraacetic acid, a compound which complexes extracellular calcium, thus preventing it from entering the cell), activated receptors that signal through the $IP_3$ system are still able to stimulate an increase in the intracellular $Ca^{2+}$ concentration. Attention then focused on the two organelle depots: either one or both could store the $Ca^{2+}$ ions mobilized by $IP_3$. Cell fractionation studies provided the answer to this question. Preparations of mitochondria and isolated vesicles of endoplasmic reticulum were loaded with radioactive $^{45}Ca^{2+}$ and the effects of $IP_3$ on the rate of release of radioactivity into the medium investigated. These studies showed conclusively that it is the endoplasmic reticulum alone that harbours the $IP_3$-sensitive calcium stores present within cells. Mitochondria play no role in the generation of the intracellular calcium signal, but they contribute to the termination of the response by soaking up excess $Ca^{2+}$ ions once they are released into the cytosol.

$IP_3$ gates intracellular $Ca^{2+}$ channels that are embedded in the membranes of the endoplasmic reticulum. The $IP_3$ receptor is thus a type of transmitter-gated ion channel (see Table 3.4), which is opened when $IP_3$ binds, presumably due to conformational changes in the structure of the channel protein. The $IP_3$ receptor/ channel has been purified and its structure characterized. It is composed of four membrane-spanning subunits, which combine to form the pore of the channel. Each subunit also has an $IP_3$ binding site at the N-terminus, which protrudes into the cytosol. The protein is homologous to another important channel involved in the mobilization of intracellular $Ca^{2+}$ stores, the ryanodine receptor (RYR), although this protein is not under the control of $IP_3$. (The mode of regulation of RYR will be discussed briefly in Section 3.9.4.)

The structure and function of purified $IP_3$ receptors have been investigated in cells and isolated membrane vesicles. These studies have revealed a number of interesting properties regarding their regulation and sensitivity to $IP_3$. In particular, the $Ca^{2+}$ concentration in the lumen of the endoplasmic reticulum appears to affect the process of $IP_3$-induced calcium release. In addition, the receptor displays a bell-shaped response curve to cytosolic calcium, as shown in Figure 3.51. This is interpreted as meaning that calcium assists $IP_3$ in the stimulation of its own release from intracellular stores up to a threshold point, then it inhibits further mobilization, thus switching off the process of $IP_3$-mediated calcium release.

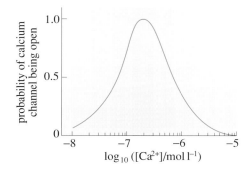

**Figure 3.51** The response of the $IP_3$ receptor to cytosolic $Ca^{2+}$ concentration.

▷ At what 'threshold' concentration of cytosolic $Ca^{2+}$ is the $IP_3$ receptor inhibited? (See Figure 3.51.)

▶ The $Ca^{2+}$ concentration coresponding to the peak of the curve, i.e. $2 \times 10^{-7}$ mol $l^{-1}$.

The sensitivity of the $IP_3$ receptor to $Ca^{2+}$ ions is a property also shared by the ryanodine family of receptor proteins. The regulatory effects of calcium play an important role in the generation of calcium oscillations and waves in many different cells (e.g. eggs, liver, endothelium) – see Plate 1.1 and Chapter 7.

A tenfold increase in cytosolic $Ca^{2+}$ concentration (from $10^{-7}$–$10^{-6}$ mol $l^{-1}$) is observed within seconds of the activation of the $IP_3$ pathway and many of the observed cellular responses are triggered, at least in part, by the release of $Ca^{2+}$ ions from the endoplasmic reticulum into the cytosol (Figure 3.52). One of the important effects of $Ca^{2+}$ ions is the coactivation of the membrane-bound form of the enzyme *protein kinase C* (PKC). You will see shortly how this process links the dual second messenger systems of $IP_3$ and DAG together.

Free cytosolic $Ca^{2+}$ ions also associates with a specific binding protein called *calmodulin*. The calmodulin polypeptide contains a predominance of acidic amino acids, which are ionized within the cell, thus generating negative charges

**Figure 3.52** The phosphatidyl inositol (PI) signalling system: the effects of increasing IP$_3$ and DAG levels in cells. (1) Ligand binds to receptor, leading to the activation of phospholipase C. (2) IP$_3$ is released into the cytosol, and binds to receptor ion channels in the endoplasmic reticulum (ER), causing a release of Ca$^{2+}$ ions. (3) The onset of the calcium signal is augmented by a process of calcium-induced calcium release. (4) The elevation in cytosolic calcium concentration produces a myriad of biological responses, including calmodulin activation (which may act on a number of downstream response elements), the stimulation of mitochondrial matrix enzymes, muscle contraction, phospholipase A$_2$ activation (leading to the production of eicosanoids – see below, and also in Chapter 4), the regulation of ion channels and PKC activity. (5) DAG stimulates PKC directly, thereby initiating intracellular protein phosphorylation cascades.

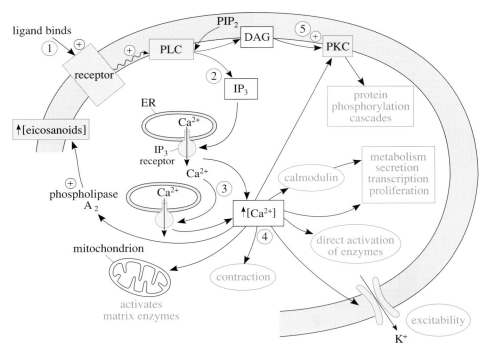

which provide a site for strong interaction with the cationic calcium ions (Figure 3.53). When the four Ca$^{2+}$ binding sites on the molecule are occupied, they induce conformational changes that favour the binding of the calmodulin/Ca$^{2+}$ complex to a number of intracellular enzymes which are subsequently activated, including calmodulin-dependent kinases. These phosphorylate a range of enzymes and other intracellular proteins which, like PKA, trigger important changes in cell function, such as altered metabolite pathways, transport and cytoskeleton rearrangement.

However, not all the effects of Ca$^{2+}$ ions require calmodulin (Figure 3.52). Other Ca$^{2+}$ binding proteins mediate muscle cell contraction. In the liver, the Ca$^{2+}$ ions released following stimulation by adrenalin (acting via $\alpha$-receptors, G$_q$ and IP$_3$) binds directly to a subunit of phosphorylase kinase. This process allosterically activates the enzyme, thus promoting the breakdown of glycogen which serves to maintain blood glucose levels (Figure 3.54). Within the mitochondrial matrix, Ca$^{2+}$ ions directly activate some of the key enzymes involved in the oxidation of pyruvate; calcium is unique in this respect since it is the only intracellular signalling species known to enter the mitochondria. Calcium can also act directly on ion channels to influence excitability and it binds to the cytosolic form of *phospholipase A$_2$*, promoting its translocation to the cell membrane. Phospholipase A$_2$ is the enzyme responsible for liberating arachidonic acid from membrane phospholipid. Arachidonic acid is the precursor of a family of local hormones (so-called because they carry information to cells in the immediate vicinity) called the *eicosanoids,* which encompasses the prostaglandins, leukotrienes and thromboxanes. (The synthesis and action of these short-range lipid mediators is discussed further in Chapter 4.)

▷ Phosphorylase kinase is also activated by cAMP via protein kinase A. What hormone activates the cAMP system in the liver?

▶ Glucagon.

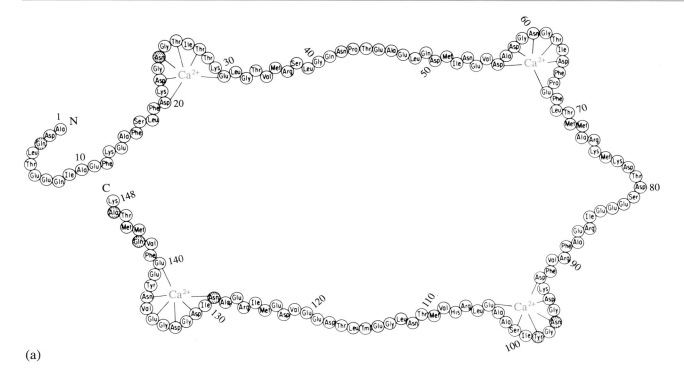

(a)

***Figure 3.53*** The structure of calmodulin. (a) Amino acid sequence. (b) Schematic diagram of three-dimensional structure. The four bound $Ca^{2+}$ ions are shown in colour.

(b)

The termination of the $IP_3$ signal is facilitated by specific phosphatases which sequentially remove phosphate groups, yielding inositol, which can then be reincorporated back into membrane phospholipid and $PIP_2$ (Figure 3.48). The action of these phosphatases is inhibited by lithium ions which, as a monovalent alkali metal, can in principle compete with either sodium or potassium. At high concentrations, lithium is toxic but low doses are used clinically to regulate the mood swings which occur in manic depression; exactly how it exerts this effect is not clear, but one possible route may be by way of its inhibitory effect on the phosphatases.

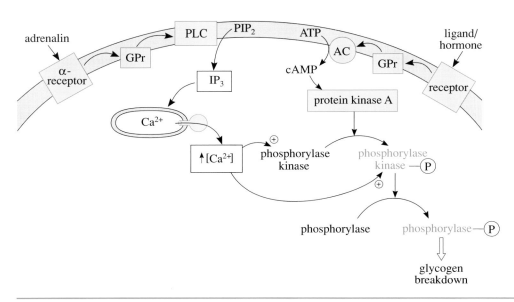

***Figure 3.54*** The role of $Ca^{2+}$ ions and cAMP in the control of glycogen breakdown in the liver. The adrenalin α-receptor is linked to phospholipase C (PLC), via a G protein (GPr). The $IP_3$ released causes an increase in cytosolic $Ca^{2+}$ concentration, which activates phosphorylase kinase, which then activates phosphorylase, resulting in stimulation of glycogen breakdown. (AC, adenylyl cyclase; PKA, protein kinase A.)

▷ How would lithium affect the inositol phosphate pathway? (Use Figure 3.48 to help you.)

▶ By inhibiting the phosphatases, the supply of inositol for the resynthesis of the pathway precursor $PIP_2$ is limited. Thus, lithium would be expected to cause a loss in the sensitivity of the phosphatidyl inositol signalling pathway in cells.

The cytosolic $Ca^{2+}$ concentration falls back to resting values within seconds following the discharge of $IP_3$-sensitive stores. It is important that the cell is not subjected to a high intracellular $Ca^{2+}$ concentration for a long period of time, since this may lead to the activation of $Ca^{2+}$-dependent proteinases, which would allow autolysis (self-digestion). This may account for the need for complex calcium spikes and waves as a long-term intracellular signalling device (discussed further in Chapter 7), rather than its prolonged elevation. The prompt decrease in cytosolic $Ca^{2+}$ concentration following the $IP_3$-mediated discharge from the endoplasmic reticulum is largely due to the activation of $Ca^{2+}$ ATPases, which respond to the elevated levels of the cation. However, only a proportion of the released calcium is taken back up into endoplasmic reticulum stores. Some cytosolic calcium is expelled across the external cell membrane (via $Ca^{2+}$ ATPases or $Na^+/Ca^{2+}$ exchange) and the remainder enters the mitochondria, driven by the negative membrane potential of the matrix (Figure 3.55). Thus, on termination of the calcium signal, the $IP_3$-sensitive calcium stores are, to some extent, left in a calcium-depleted state. Clearly, the cell must have some mechanism for 'topping up' the calcium content of its endoplasmic reticulum, otherwise $IP_3$-mediated responses would soon diminish (and this only occurs when very high doses of agonist are applied to cells).

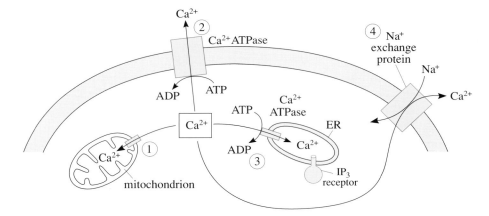

**Figure 3.55** Mechanisms controlling cytosolic free $Ca^{2+}$ concentration. The figure shows the routes by which levels decrease following $IP_3$ stimulation: there is a movement of $Ca^{2+}$ ions from the cytosol into the mitochondrial matrix (1) and the lumen of the endoplasmic reticulum (3) and an efflux from the cell via the cell membrane-bound $Ca^{2+}$ ATPase (2) and the $Na^+/Ca^{2+}$ exchange protein (4).

To understand how cells manage to charge up these endoplasmic reticulum calcium reserves we must return to some experimental observations discussed earlier. Responses to ligands that signal via the $IP_3$ system are not immediately affected by the presence of substances (like EGTA) that sequester extracellular calcium (Figure 3.56).

▷ How can EGTA interact with $Ca^{2+}$ ions? (Figure 3.56)

▶ Its negative charges interact with the cation, allowing it to act as a chelating (binding) agent.

However it appears that such responses cannot be sustained if cells are incubated for longer periods in EGTA-containing medium. This shows that extracellular calcium is important for the maintenance of an effective $IP_3$ signalling pathway.

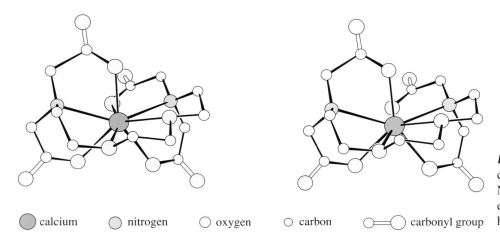

| ⬤ calcium | ◯ nitrogen | ◯ oxygen | ◯ carbon | ◯=◯ carbonyl group |
|---|---|---|---|---|

**Figure 3.56** Stereoviews of the calcium–EGTA complex, $[Ca(EGTA)]^{2-}$. Note that the bonds to the $Ca^{2+}$ ion denote electrostatic interactions. (For clarity the hydrogen atoms have been omitted).

In fact, extracellular sources are used to boost $Ca^{2+}$ levels in the endoplasmic reticulum following $IP_3$ stimulation. $Ca^{2+}$ ions are thought to enter the cell through a calcium release activated channel. Interestingly, the influx of external $Ca^{2+}$ appears to be gated by some mechanism that is sensitive to the emptying of the ER stores so the channel functions rather like a capacitor, drawing on extracellular calcium reserves whenever the intracellular stores begin to dwindle. There is no clear explanation as to how capacitative $Ca^{2+}$ entry may operate, but two models have been put forward (Figure 3.57). The first model suggests that a calcium influx factor (CIF), which is as yet unidentified, is released from the ER as the lumenal calcium reserves diminish and this triggers the opening of $Ca^{2+}$ channels in the cell membrane. The second model proposes that a conformational coupling system exists, so that the depleted ER calcium store induces a change in the conformation of the $IP_3$ receptor which results in an association with and a subsequent opening of the cell membrane $Ca^{2+}$ channel.

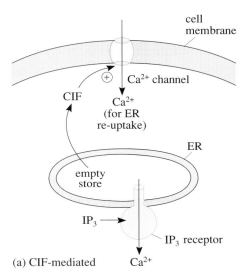

(a) CIF-mediated

### Diacylglycerol (DAG) and the activation of protein kinase C

Within the confines of the cell membrane, another signalling pathway is in operation. DAG binds to and activates protein kinase C (Figure 3.52), a multifunctional enzyme which regulates a wide variety of cellular events, from the phosphorylation of enzymes involved in glycogen metabolism (in liver) to very sophisticated responses such as cell growth and proliferation (mitogenesis) and even possibly the changes in connections between nerve cells which occur during learning and memory formation. However, the activation of PKC is not only dependent on DAG. The enzyme also requires phosphatidyl serine (another membrane component) and $Ca^{2+}$ ions before the stimulatory effect of DAG is manifest. Since the intracellular $Ca^{2+}$ concentration is increased by $IP_3$, you can appreciate that the activation of PKC is a coordinated event involving both limbs of the PI signalling pathway (Figure 3.52). Note that DAG (like $IP_3$) is also recycled back into phosphatidyl inositol (via phosphatidate) and then $PIP_2$, thus preventing the exhaustion of pathway precursor (Figure 3.48).

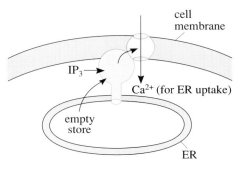

(b) via conformational coupling

**Figure 3.57** Models to explain capacitative $Ca^{2+}$ entry into cells for recharging ER calcium stores. (a) Entry mediated by a calcium influx factor (CIF), released from the ER as lumen stores diminish. (b) Conformation coupling. The empty ER store induces a change in the conformation of the $IP_3$ receptor, which is transmitted to a $Ca^{2+}$ channel in the cell membrane.

### Biological responses require the activation of both limbs of the $IP_3$/DAG pathway

Both the transient elevation in $Ca^{2+}$ concentration (mediated by $IP_3$) and the stimulation of PKC (by DAG) appear to be obligatory requirements for many of the biological responses elicited via the PI signalling pathway. Often the outcomes of the two limbs of the pathway complement each other. For example, the stimulation of α-receptors in liver cells by adrenalin promotes the breakdown of glycogen in two ways:

1   $Ca^{2+}$ stimulates phosphorylase kinase, leading to the activation of phosphorylase (Figure 3.54).

2   PKC phosphorylates glycogen synthetase and inactivates it.

Thus, glycogen breakdown is stimulated, while the synthesis pathway is inhibited. Similarly, the process of cell proliferation requires two key intracellular events: an increase in cytosolic $Ca^{2+}$ concentration and an increase in pH (alkalinization). The calcium signal is, of course, mediated by $IP_3$. Alkalinization of the cytosol is promoted by PKC, by way of the phosphorylation and activation of an $Na^+/H^+$ antiporter (Figure 3.58).

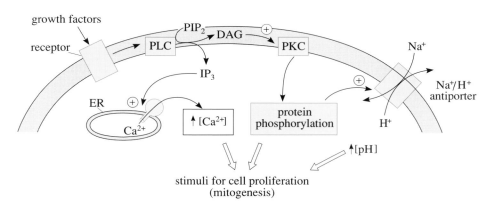

**Figure 3.58**   Phospholipase C activation stimulates cell proliferation via both limbs of the PI signalling ($IP_3$/DAG) pathway.

One experimental approach clearly illustrates the physiological importance of synergy between the two limbs of the pathway. The effects of $IP_3$ can be mimicked by the use of a *calcium ionophore* (A23187).

▷   What is the role of A23187?

▶   An ionophore is an ion carrier (Chapter 2, Section 2.5). Thus A23187 allows $Ca^{2+}$ ions to cross the cell membrane and so increases the cytosolic $Ca^{2+}$ concentration.

PKC can be stimulated independently of DAG by phorbol esters: these are substances that resemble DAG and so also allosterically activate the enzyme.

The effects of A23187, phorbol esters and the two in combination have been investigated in a number of different cell types. It appears that both are needed for lymphocyte proliferation and also for the secretion and aggregation of platelets (which is an important event in blood clotting). Thus, both $Ca^{2+}$ and PKC contribute to these cellular responses which are coordinated through the ligand activation of phospholipase C.

### Temporal aspects of the $IP_3$/DAG pathway

PKC activation is also important in regulating the termination of the PI signalling pathway. It provides a route for feedback inhibition by phosphorylating phospholipase C and thus causing a decrease in its activity. As a result of this and the termination events associated with the $IP_3$/calcium signal discussed earlier, many of the effects of the PI signal transduction pathway are immediate and short-lived. However, some longer-term effects (e.g. cell proliferation, fertilization) have also been identified and these are thought to be the result of the chronic activation of PKC. The initial activation of PKC (via the stimulation of phospholipase C) can act as a trigger for a second (delayed) phase of response which is characterized by the sustained generation of DAG in the cell membrane. Long-term effects are due to the positive feedback of PKC on *phospholipase D*, which when stimulated provides a secondary source of DAG to prolong the activation of PKC (Figure 3.59a). Another phospholipase C (cf. the phospholipase C that acts on phosphatidyl inositol), which acts on phosphatidyl choline in the cell membrane, also contributes to the second phase of DAG release since this enzyme is also stimulated by PKC (Figure 3.59b).

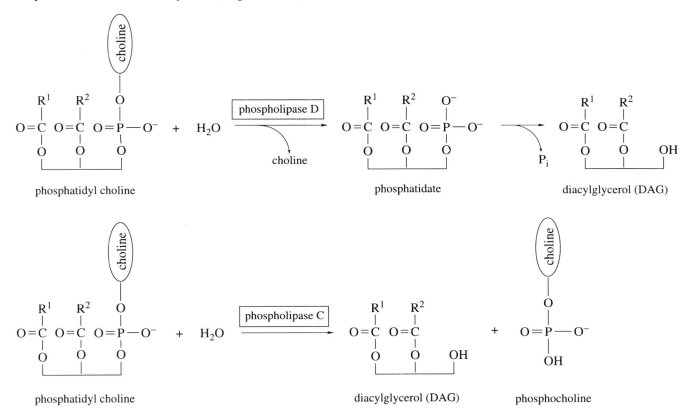

### 3.9.4   Calcium effects that do not involve phospholipase C

In the calcium responses we have discussed so far, $Ca^{2+}$ ions are released from intracellular stores in response to an increase in the concentration of the intracellular second messenger $IP_3$. However, there are other ways in which cytosolic $Ca^{2+}$ levels can be increased and these processes also play a role in intracellular signalling. **Voltage-gated calcium channels** in the cell membrane allow extracellular $Ca^{2+}$ ions to access the cytosol (entering down their concentration gradient) and act as a second messenger. Many of the biological responses are similar to those observed in $IP_3$ systems and often involve calmodulin. Physiologically important voltage-gated $Ca^{2+}$ channels are found in cardiac muscle. The resultant rise in cytosolic $Ca^{2+}$ level (due to membrane depolarization) triggers an explosive release of $Ca^{2+}$ ions from the sarcoplasmic reticulum (SR); this process of calcium-induced calcium release is essential for every heart beat (Figure 3.60a). At nerve synapses, $Ca^{2+}$ entry acts as a stimulus for the release of neurotransmitters. In skeletal muscle, ryanodine receptors (RYR) located in the sarcoplasmic reticulum are activated by a dihydropyridine receptor (DHPR) in the cell membrane. The DHPR reacts to a change in voltage at the cell surface and undergoes a conformational change which is physically transmitted through to

**Figure 3.59** The degradation of phosphatidyl choline by phospholipases. (a) The action of phospholipase D. The phosphatidate produced in this reaction is further degraded to DAG. (b) The action of phospholipase C, producing DAG directly.

the RYR, which then opens, releasing the stores of calcium (Figure 3.60b). The major effect of $Ca^{2+}$ ions in this tissue is, of course, the stimulation of muscle contraction, although, as we have seen, calmodulin and enzymes such as phosphorylase kinase are also activated.

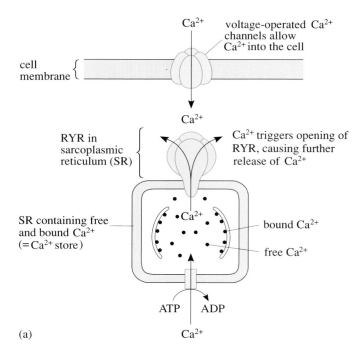

(a)

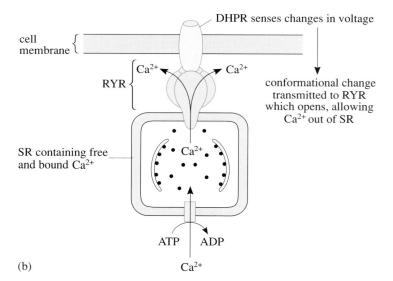

(b)

**Figure 3.60** (a) Calcium signalling in the heart. (b) Calcium signalling in skeletal muscle.

### 3.9.5 Tyrosine kinases and signal transduction

You will recall from an earlier discussion that autophosphorylation is a distinguishing feature of receptors that possess intrinsic tyrosine kinase activity and this is instigated by the binding of ligands, typically growth factors such as platelet-derived growth factor (PDGF). The organization of the downstream transduction systems is complex and the pathways from the receptor to the final responses (which ultimately involve the modulation of gene expression) are only just beginning to be worked out.

Some important discoveries have been made about the proximal (receptor) section of the pathway, though, and these are highlighted in Figure 3.61, which shows the transfer of information from the cell surface into the cytosol and nucleus (ultimately causing cell proliferation) following the binding of PDGF to its receptor.

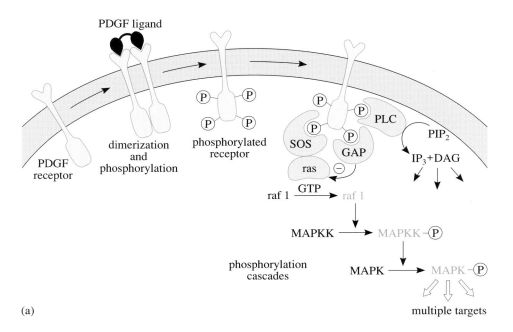

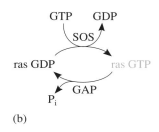

(a)

(b)

**Figure 3.61** Signalling through the platelet derived growh factor (PDGF) receptor. (a) Summary of the whole pathway. (b) Details of ras activation. SOS is a GTP/GDP exchange protein which is activated by the PDGF receptor; ras is the signalling component which is activated when its bound GDP is replaced by GTP (via SOS); GAP is a GTPase activating protein involved in terminating the actions of ras; raf 1 is a protein kinase which initiates the phosphorylation cascades. The multiple targets of these cascades include the phosphorylation and activation of transcription factors which regulate genes controlling proliferation. MAPKK, mitogen-activated protein kinase kinase; MAPK, mitogen-activated protein kinase. (Note that phospholipase C is also activated via the PDGF receptor.)

The signalling sequence begins with PDGF inducing dimerization of two PDGF receptors. Transduction is facilitated by each of the receptor tyrosine kinase domains phosphorylating each other on specific tyrosine residues. The phosphorylated tyrosine residues then act as docking sites, binding different amplifiers (SOS, GAP, phospholipase C) which are present in the cytosol. The result of receptor activation is a multimolecular complex which triggers a number of signalling pathways, including the stimulation of phospholipase C (Figure 3.61a). A route which appears to be important for initiating mitosis involves a GTP/GDP exchange protein called **ras**. When bound to GDP ras is inactive, but the protein can be activated by SOS, one of the amplifiers that associates with the phosphorylated receptor. SOS promotes the exchange of GDP for GTP on ras (Figure 3.61b) so, rather like the $\alpha$ subunit of a G protein, GTP-bound ras is then able to transmit information to the next component in the pathway, a protein kinase called raf 1. There is another similarity between ras and the G protein $\alpha$ subunit: ras is able to inactivate itself due to intrinsic GTPase activity, although in this case the 'switch-off' mechanism also requires the assistance of GAP, a GTPase activating protein (Figure 3.61). This complex regulation of ras activity is crucial to the control of cell proliferation.

Downstream from ras, phosphorylation cascades become operational. The key proteins involved in this section of the pathway include a number of *mitogen-activated protein kinases* (MAPKs) – not to be confused with MAP (microtubule-associated protein), discussed in Book 1. These kinases, which are specific for serine and threonine residues, are ubiquitously present in cells and play a critical role in cell proliferation processes. There are literally thousands of proteins that may be targeted by MAPKs, including nuclear proteins which control cell cycle events, although many of the details are still unknown.

### 3.9.6   Interactions between second messenger systems

You should now appreciate that all cells are furnished with a number of different second messenger systems which permit communication with the external environment. Cells vary in their signalling repertoire, but rarely employ only one mode of transduction. There exists, therefore, tremendous scope for signal interactions and it is the interplay between these different systems that permits such fine control over biological processes. 'Cross-talk' may occur on a number of different levels: G proteins, amplifiers or at the actual target proteins (response elements) which generate the biological responses. In the latter situation, signal transduction pathways may converge on similar intracellular pathways, so that two stimuli may synergize and produce an enhanced biological response.

▷   Recall the mobilization of glycogen in liver. What are the two signalling pathways which activate this process, and how do they interact (Figure 3.54)?

▶   cAMP (through the action of glucagon) and $IP_3$/DAG (through the action of adrenalin on $\alpha$-receptors) stimulate glycogen breakdown in liver. Phosphorylase kinase is activated by PKA *and* $Ca^{2+}$, leading to the activation of phosphorylase. In a coordinated manner, glycogen synthetase is phosphorylated and inhibited by PKA and PKC.

In contrast, some convergent second messenger systems oppose each other and produce antagonistic responses. For example, $Ca^{2+}$ promotes the contraction of smooth muscle, whereas cGMP mediates relaxation.

### Summary of Section 3.9

Cells employ second messenger systems as a means of amplifying extracellular signals into intracellular responses. The second messengers we have discussed in this section all involve enzymes that are regulated by the binding of ligands to specific receptors. In turn, the second messenger molecules generate a cascade of intracellular events which culminate in particular cellular events. As with all signal transduction components, inactivation systems operate in order to maintain a cell's sensitivity to its environment. Signal termination is facilitated by additional enzymes or transport systems which promote the elimination of second messenger molecules from the cell.

The most well-established amplifier is the enzyme adenylyl cyclase, which is located in the cell membrane and is under the control of G proteins. Its product, cAMP, instigates multiple phosphorylation cascades via the activation of protein kinase A. The ultimate cellular response depends on the target tissue: in muscle, the cAMP cascade stimulates glycogen breakdown while inhibiting the corresponding synthetic pathway, whereas in adipose tissue the same receptor ($\beta$-adrenalin) stimulates lipolysis.

Ligand binding to either specific G protein-linked or tyrosine kinase receptors can activate another important effector enzyme, phospholipase C (PLC). This enzyme generates two second messengers, $IP_3$ and DAG, which operate by distinct signalling routes, via $Ca^{2+}$ and protein kinase C (PKC) respectively. PKC, like protein kinase A, causes protein phosphorylations and $Ca^{2+}$ ions, via calmodulin, can also initiate similar kinase cascades. In addition, calcium has a number of other intracellular effects including direct enzyme activation and the production of local hormones, such as the prostaglandins and nitric oxide. Most PLC responses involve interactions between the two limbs of the pathway. The $IP_3$-mediated mobilization of $Ca^{2+}$ from intracellular stores (endoplasmic reticulum/

sarcoplasmic reticulum) often generates complex patterns of calcium waves. DAG levels in the cell membrane may be sustained beyond the time of ligand activation, and these processes are thought to contribute to longer-term signalling which is necessary for controlling cell proliferation and differentiation. Calcium can also function as a second messenger independently of the $IP_3$ system. For example, in skeletal muscle, the release of $Ca^{2+}$ ions from the sarcoplasmic reticulum is achieved via the dihydropyridine/ryanodine receptors, which respond to a change in voltage across the cell membrane.

In the case of receptor tyrosine kinases (such as the PDGF receptor), the downstream events which follow ligand activation are not well-established. The sequence of events involving some of the 'early' components of TK signalling pathways have been identified, including the binding of cytosolic amplifier proteins such as SOS, PLC and GAP, the GTP/GDP exchange protein ras and MAPKs, but the distal region of the signalling pathway (i.e. that furthest from the receptor), which is likely to involve multiple gene regulation, has not yet been dissected.

## 3.10  Concluding overview

A number of interactive transduction systems generate a diverse spectrum of biological responses in cells following activation by extracellular stimuli. A summary of the major signal processing pathways for chemical stimuli is shown in Figure 3.62. Transduction is often associated with signal amplification (for example, through the production of second messengers) and in some cases, also with the maintenance of signalling pathways over an extended period of time. Termination mechanisms operate at all levels, ensuring a cell's continued responsiveness to stimuli. Knowledge of these transduction pathways offers the

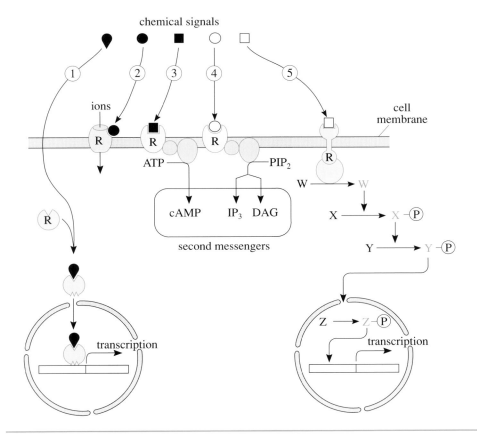

***Figure 3.62*** A summary of the major signal processing pathways activated by chemical stimuli: (1) steroid hormones use cytosolic receptors to gain rapid access to the nucleus; (2) transmitter-gated receptors, used primarily by neurotransmitters, induce rapid changes in membrane potential by gating ions; (3 and 4) G protein-linked receptors generate diffusible second messengers that can rapidly disperse information throughout the cell; (5) tyrosine kinase-linked receptors employ a protein phosphorylation cascade (X to Z) to transmit information into the nucleus.

potential for the pharmacological intervention into the functioning of various signalling components: this ability to manipulate the sensitivity of cells to the environment may yield therapeutic benefits for the treatment of a number of diseases.

## Objectives for Chapter 3

After completing this chapter, you should be able to:

3.1   Define and use, or recognize definitions or applications of, each of the terms printed in **bold** text.

3.2   Describe the general organization of transporter proteins and explain how they are able to permit lipid-impermeable substances to cross the phospholipid bilayer.

3.3   Distinguish between simple and carrier-mediated (facilitated) diffusion from knowledge of transport kinetics, and describe why binding proteins and chemical modification enhance rates of diffusion across the cell membrane.

3.4   Show how the sodium/potassium pump uses energy to move substances against concentration gradients.

3.5   Identify the characteristics of carriers and channels and use these to distinguish between them.

3.6   Identify the key components of intracellular signalling systems and describe how they transmit information.

3.7   Suggest why transmitter-gated channels instigate more rapid responses than those operating via G proteins, and why the responses generated by hydrophobic ligands take longer than those of hydrophilic ligands.

3.8   Provide examples illustrating the operation of transmitter (ligand)-gated and voltage-gated channels.

3.9   Describe the sequence of events involved in G protein signal transduction, and apply this knowledge to a range of G protein-linked systems.

3.10   Describe the upstream and downstream events surrounding the the production of second messengers such as cAMP and $IP_3$.

3.11   Provide an example of at least one phosphorylation signalling cascade, showing how kinases and phosphatases regulate cellular processes.

3.12   Give examples showing that receptors, G proteins and second messenger systems can be artificially activated or inhibited by compounds which can be experimentally and/or clinically useful.

## Questions for Chapter 3

### Question 3.1   (Objectives 3.1 and 3.3)

Compare and contrast the processes of diffusion and facilitated transport into cells. How might these two modes of transport be enhanced? What factors might influence whether a substance is transported by either of these two routes?

### Question 3.2   (Objective 3.2)

If substance Z is transported into a cell by active transport, which of the following criteria must be satisfied?

(a)  The concentration of Z outside the cell must be higher than that inside.

(b)  The cell membrane must contain a membrane-spanning protein that interacts with Z at the external surface of the cell.

(c)  The cell must have functional energy-generating systems.

(d)  Binding proteins for Z must be located in the cytosol.

(e)  Z must be hydrophobic.

### Question 3.3   (Objective 3.2)

Transporter proteins are amphipathic molecules, i.e. they contain hydrophobic and hydrophilic regions. Why should this always be the case?

### Question 3.4   (Objective 3.4)

Using genetic engineering techniques, a modified form of $Na^+/K^+$ ATPase was produced in which an aspartate residue on the $\alpha$ subunit was replaced with a glycine residue. The mutated protein was introduced into phospholipid vesicles but failed to function, even when ATP was provided in both the intra- and extravesicular medium. Can you explain these observations?

### Question 3.5   (Objectives 3.1, 3.6 and 3.7)

Classify each of the following statements as true or false, and if they are false say why.

(a)  All signal transduction processes that occur at the cell membrane involve G proteins.

(b)  A transmitter-gated ion channel functions as a receptor and as an effector.

(c)  Some ligands interact with more than one type of receptor.

(d)  Signal transduction is more rapid through the muscarinic acetylcholine receptor than through the nicotinic acetylcholine receptor.

(e)  All signal transduction processes occur at the cell membrane.

(f)  Chemical ligands form covalent bonds with the receptors that they activate.

### Question 3.6   (Objectives 3.1, 3.8 and 3.9)

Where are you most likely to find the following? (a) Nicotinic acetylcholine receptor (transmitter-gated); (b) transducin; (c) voltage-gated ion channels.

*Question 3.7  (Objective 3.9)*

(a) Place the following events on Figure 3.63 to construct a G-protein-linked signal transduction sequence:

GTP hydrolysed to GDP and G protein components reassociate

Membrane-bound enzyme generates second messenger

Agonist binds to receptor

GTPα activates amplifier

G protein binds GTP and dissociates into two components

(b) How do G proteins facilitate signal amplification?

(c) How is the signal terminated?

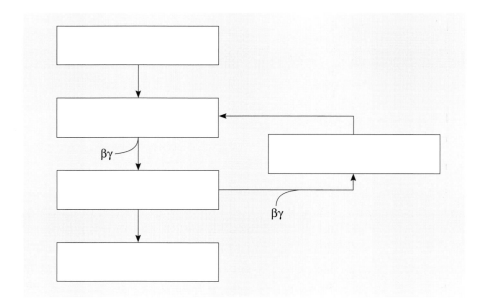

***Figure 3.63***  For Question 3.7a.

*Question 3.8  (Objectives 3.9 and 3.10)*

(a)  You suspect that hormone Z produces its biological response in target cells via a G protein. How could you test this hypothesis?

(b)  Would it be possible to identify the type of G protein involved in the response?

(c)  In order to confirm the involvement of G proteins, a second experiment was conducted. Target cells were incubated with $[AlF_4]^-$, an ion that associates with inactive GDP-bound forms of G proteins and mimics a terminal phosphate group, so that the G protein appears to bind GTP. What would you expect to observe in such an experiment?

(d)  The effects of hormone Z could also be mimicked by forskolin. Can you suggest the G protein, amplifier and second messenger produced in this signalling pathway?

*Question 3.9  (Objectives 3.6 and 3.9)*

(a)  Identify the targets of the following G proteins from the list of amplifiers given below: (i) $G_i$; (ii) $G_s$; (iii) transducin, $G_t$; (iv) $G_q$.

(b) Which of the amplifiers listed are **not** G protein-linked? How are these enzymes activated?

*Amplifiers*: tyrosine kinase; adenylyl cyclase; cGMP phosphodiesterase; phospholipase C; guanylyl cyclase.

## Question 3.10   (Objective 3.10)

Which of the following events could be associated with the activation of phospholipase C?

(a)  Phosphorylation of a tyrosine kinase receptor

(b)  Increase in cytosolic $IP_3$ concentration

(c)  Activation of the dihydropyridine receptor

(d)  Mobilization of calcium from intracellular stores

(e)  Formation of cAMP

(f)  Serine/threonine phosphorylation of intracellular proteins

(g)  Increase in phospholipase $A_2$ activity

(h)  Increase in protein kinase C activity

(i)  Ligand-mediated activation of $G_q$

(j)  Increase in diacylglycerol (DAG) levels in the cell membrane

(k)  Increase in protein kinase A activity

(l)  Activation of calmodulin

Draw a flow diagram to show the relationships between the relevant processes.

## Question 3.11   (Objective 3.12)

An experiment was carried out to investigate the ability of acetylcholine and the related substances propanoyl and hexanoyl choline to produce contraction in isolated rat smooth muscle.

(a)  From the data in Table 3.9, which of these two choline derivatives is the more potent?

(b)  Can you explain the altered $pD_2$ value for acetylcholine in the presence of hexanoyl choline?

**Table 3.9**  For Question 3.11.

|  | $pD_2$ |
| --- | --- |
| acetylcholine | 7.0 |
| propanoyl choline | 5.0 |
| acetylcholine plus 1 $\mu$mol $l^{-1}$ hexanoyl choline | 4.8 |

## CHAPTER 4

**Intercellular communication**

## 4.1  Introduction

The previous chapter reviewed the biochemical mechanisms involved in the transfer of information inside cells and focused particularly on the events that occur as a signal is transmitted across the external cell membrane to the interior of the cell. In this chapter, we look at the cell membrane from a different perspective, viewing it from the outside as the principal communication 'switchboard' in multicellular organisms. The evolution of intercellular communication systems has accompanied the evolution of multicellularity. Chapter 1 of this book made the point that multicellular organisms, by virtue of their size and the specialization of their cells, have evolved mechanisms of communication, coordination and control between cells that can operate over a huge range of distances and time frames, from micrometres and milliseconds to metres and months, or even years. Coordination and control require the reliable transmission of information to specific target cells at the appropriate time. Here, and in the following chapter, we examine cell membranes as receptive surfaces for the multiple stimuli arriving from the internal and external world and describe three interacting physiological systems – the endocrine, nervous and immune systems – which generate many of these signals in multicellular animals. In this chapter we focus almost entirely on animal physiology, particularly that of mammals. We also review a variety of very short-range communication mechanisms that cannot easily be categorized as originating in one of these systems, but which have a vital role in the transfer of information between cells. Finally we look at communication across the fixed 'gaps' between adjacent cells in tissues of various kinds.

All vertebrate animals and most invertebrates have communication systems that share important features with the endocrine, nervous and immune systems of mammals. (There is no equivalent of the nervous or immune systems in plants; there are plant hormones but these are not discussed in this chapter.) These three physiological systems have evolved to a high degree of complexity in mammals and it is not our intention to describe any of them in detail. Rather, our aim is to select certain key features of these systems to illustrate the variety of means by which cells communicate with each other across the available range of time and space, starting with the transmission of long-range or sustained information and ending with short-range and short-lived signalling mechanisms. It won't be possible to keep strictly to this shift of scale from longer to shorter as the chapter proceeds (biological systems are never that neat), but this is the general framework we have in mind.

Before looking at particular mechanisms of cell-to-cell communication in detail, it is useful to have some idea of the 'areas of expertise' of the three physiological systems discussed in this chapter. The **endocrine system** is the collective term given to certain glands in the body, each composed of specialized cells, and the signalling molecules that they synthesize and secrete. These molecules are known as *endocrine hormones* (often abbreviated to 'hormones' but, as you will see later, another group of signalling molecules, which interact either with the very cells that produce them or with those nearby, is referred to as 'local hormones' – there are two types *autocrines* and *paracrines* (Section 4.4) – so we will generally use 'endocrine hormones' to avoid confusion). Central to the function of the endocrine system is the fact that each endocrine gland is richly supplied with blood capillaries; the hormone is secreted from the cells of the gland directly into the extracellular fluid, it diffuses across the capillary wall into the bloodstream and is transported around the body. Once released from the gland, an endocrine hormone becomes *systemically* distributed, that is it circulates throughout the whole of the vascular system and so comes into contact with all parts of the body.

▷ Does this mean that endocrine hormones *affect* all parts of the body?

► No. A hormone can only have an effect on target cells that express appropriate receptors which bind the hormone.

The receptors for some hormones are indeed widely distributed on many different types of cell in the body (an example is the hormone *insulin* which is discussed later), whereas others have extremely limited receptor distribution and hence the hormone only affects a specific category of target cells. The endocrine system thus has the potential to affect many tissues and organs simultaneously, as well as to deliver signals with pin-point precision to a single kind of target. Since endocrine hormones are generally synthesized and stored in the cells of the gland before release, they can be secreted into the bloodstream either in a large 'burst', or over a long period of time, so a sustained signal can be generated with long-term effects on the target tissues. As you might expect from this property, hormones are involved in initiating and sustaining growth and development of the body and in regulating metabolism throughout life.

▷ Can you see any limitations inherent in a signalling system that relies on the bloodstream for transportation across long distances?

► It is quite slow and relatively large quantities of each hormone have to be produced in order to achieve an effective concentration in the whole of the blood volume.

In an adult human at rest, a volume of blood equivalent to the total contents of the bloodstream is pumped through the heart every minute, so it takes at least a minute to transport hormones to the most distant targets. The decreased blood flow to (for example) muscles and skin during digestion, can mean a longer delay in hormones reaching those organs and tissues. However, some endocrine glands are physically very close to their principal target tissue and this reduces the time lag; for example, the hormone *testosterone* is produced by endocrine cells in the testes and some of its effects are on sperm production in the same organ.

Another feature that endocrine glands share is that all of them are *innervated*, that is they are penetrated by nerve endings and receive electrical signals from the **nervous system**. Electrical signalling is one of the control mechanisms that regulate secretion of endocrine hormones from the originating gland. In comparison with the endocrine system, the nervous system can carry signals across long distances extremely fast. Nerves may be a metre or more in length (much longer in giraffes!) and can conduct signals at speeds of between 1 and 100 m s$^{-1}$. The nervous system delivers messages with high precision and specificity, because the signal travelling along any particular nerve pathway is transported only to those target cells with which the nerve endings make a specific interface known as a *synapse*. The electrical signal arriving at the synapse triggers the release of a chemical signalling molecule, called a *neurotransmitter*, which carries the signal across the tiny gap between the nerve ending and the target cell. We will discuss synapses and neurotransmitters in some detail at the end of this chapter as a prelude to Chapter 5, which focuses on the biochemical mechanisms underlying the generation of electrical signals in the nervous system.

The kinds of information signalled by the nervous system to target cells elsewhere in the body could be characterized as being about rapidly changing events, which require moment-to-moment updating. By altering the pattern of electrical stimulation delivered to a target cell, the nervous system is capable of exquisite 'fine tuning' of the target cell's response.

▷ Why can't the endocrine system readily cope with communicating rapidly changing information?

▶ Because once a hormone has been released into the bloodstream it cannot be 'recalled'; it continues to circulate for minutes or hours until its chemical structure begins to break down or it is excreted.

Both the nervous and the endocrine systems are engaged in signalling information about the internal state of the body and triggering response mechanisms which keep all the physiological parameters (such as temperature, blood sugar levels, hydration, etc.) close to the optimum for efficient functioning. This moment-by-moment activity is that of maintaining *homeostasis*; over the longer period of the lifetime of an organism, the process of responding to both the internal dynamics of development and the external pressures of the environment is termed *homeodynamics*. Thus the nervous system, via its sensory organs, is continuously collecting and interpreting information about the outside world and about the 'position' of the body relative to those external stimuli so that, for example, evasive action can be taken if a threat to the body's integrity is perceived.

The third physiological system introduced in this chapter is also adapted to protect the organism from external threats of a certain kind. The **immune system** is a network of cells and molecules distributed throughout the body, with the specific function of signalling information about the inappropriate entry of living or inert matter. In common speech, the immune system is often described as being on surveillance for 'foreign bodies', but in practice the stimuli it is sensitive to – at least in some individuals – range from infectious organisms such as bacteria and viruses, to proteins generated by other species (e.g. fragments of pollen or cat fur, the toxins produced by certain bacteria), debris such as grit or splinters, and industrial chemicals or drugs like penicillin. The various cells involved in the immune response to these external stimuli are derived from so-called *stem cells* in the bone marrow; their precise nature need not concern us here, but note that it is central to the function of the immune system as a communication network that huge numbers of immunologically active cells can be found in almost every part of the body.

Wherever infectious organisms penetrate the body's external barriers, they will be rapidly detected and responded to by the immune system. Chemical signals are generated by the cells that arrive first on the scene, triggering many different response mechanisms, including the attraction of more cells with a defensive role. Thus, the response is initially rapid and highly localized, but is then 'scaled up' in intensity and range. This is essential because an infection that has penetrated the body in one place may have got in elsewhere and can in any case be distributed to distant sites via the bloodstream. The immune system has evolved 'layers' of defence mechanisms, which spread outwards from the initial site of infection, like the ripples from a pebble dropped into a pool. Some of these mechanisms take days to develop and can be sustained for weeks or months. Once a certain infection has been eliminated by these long-range mechanisms, the immune system undergoes permanent changes which enable it to respond far more quickly if the same infection is ever encountered again.

▷ How has this property of the immune system been exploited by medical science to protect infants from common but potentially fatal infections?

▶ Vaccination introduces harmless killed or attenuated infectious organisms (or their products) into the body, which is stimulated to produce an immune response against them. In the process, the immune system adapts so that subsequent infection with the live organisms triggers a fast and (hopefully) protective immune response.

Thus, the immune system is capable of relatively short-term responses anywhere in the body and also gives a measure of lifelong protection against commonly encountered infections. Like the endocrine and the nervous systems, it can operate over a wide range of time and space, and communicates with many target cells of different specializations in the process of coordinating an effective response.

The three physiological systems we have sketched here are highly interactive; each of them communicates with cells in the other. For example, the nervous system fine-tunes the responses of both the endocrine and immune systems and is in turn affected by the molecules they secrete and by the homeostatic and homeodynamic state of the body which they help to regulate. Some of the signalling molecules involved in cell-to-cell communication are common to more than one system: for example, the endocrine hormone *adrenalin* is also a neurotransmitter substance carrying signals across certain synapses in the nervous system. In the rest of this chapter, we focus on examples of intercellular communication drawn from these three systems and briefly review some other important short-range mechanisms that cannot be so neatly categorized.

## Summary of Section 4.1

The evolution of multicellularity has been accompanied by the evolution of intercellular communication mechanisms, which facilitate coordination and control of organized structures formed from many different cell types, each with specialized functions. In mammals, three major communication systems can be identified, which have features in common with similar systems in other vertebrates and in some invertebrates.

The endocrine system is composed of glands that secrete long-range chemical signalling molecules, called endocrine hormones, into the bloodstream for transport to target tissues elsewhere in the body. This method of communicating information is relatively slow (minutes to hours), but can be sustained over long periods (months to years in some cases), and can simultaneously reach widely dispersed targets in many different tissues as well as highly localized and specialized targets. Many kinds of plants also use long-range chemical signalling via molecules referred to as plant hormones.

The nervous system carries information in the form of electrical impulses at extremely high speed over the full range of possible distances in the body (micrometres, to metres in large animals) and delivers the signal with exquisite precision to single target cells; signals are translated into chemical messenger molecules (neurotransmitters) at synapses.

The immune system detects the presence of inappropriate cells and molecules in the body, particularly those associated with infectious organisms; this elicits a series of communication events between many different cell types, resulting in a coordinated and precisely targeted attack at the site of infection which spreads to give protection against that infection throughout the body and, in some cases, for the rest of the organism's life.

Multicellular organisms have also evolved a number of short-range communication mechanisms that carry information between cells, which are not unique to any of the three systems summarized above.

## 4.2 Endocrine hormones: long-range chemical signals

Figure 4.1 shows the major endocrine glands which are common to mammals (though shown here in their location in humans). Examples of the hormones that they synthesize and secrete, along with their chemical properties and physiological functions, are outlined in Table 4.1. Many of these hormones were originally identified by experiments which identified the physiological effects of removing the gland from experimental animals. The symptoms of hormone deficiency could then usually be reversed by re-implantation of the gland or injection of an extract of gland tissue.

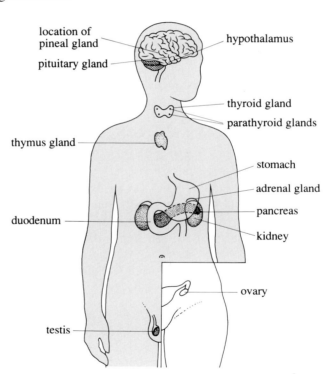

**Figure 4.1** Locations of major endocrine glands in humans; corresponding glands can be found in all mammals.

Endocrine hormones encompass a broad class of compounds, with a multitude of physiological and biochemical effects, as Table 4.1 reveals. We will return to this table several times in this part of the chapter, although we will not be discussing specific hormones in detail (this would take us further in the direction of physiology and pathology than is relevant for this Course). However, we can identify some general principles of hormone structure and relate this to their mode of action.

### 4.2.1 The chemical structure of endocrine hormones

The hormones listed in Table 4.1 can be conveniently grouped into two categories, depending on their ability to penetrate the cell membrane.

**Table 4.1** Some examples of endocrine hormones: release, mode of transport and biological effects.

(a) Peptides/polypeptides

| Hormone | Gland | Major factor(s) regulating hormone secretion | Chemical structure | Mode of transport | Target tissue(s) and some important biological effects |
|---|---|---|---|---|---|
| thyroid stimulating hormone (TSH) | pituitary (anterior lobe) | hypothalamus – thyrotropin releasing factor (TRF) stimulates | glycoprotein – two peptide chains | free in plasma | *thyroid gland* – stimulates release of thyroid hormones |
| growth hormone (GH) | pituitary (anterior lobe) | hypothalamus – GH-releasing factor stimulates and somatostatin inhibits | single peptide chain | free in plasma | *liver* – stimulates release of growth factors (somatomedins) *muscle* – stimulates protein synthesis *adipose* – stimulates lipolysis |
| adrenocortico-tropic hormone (ACTH) | pituitary (anterior lobe) | hypothalamus – corticotropin-releasing factor stimulates (CRF) | single peptide chain | free in plasma | *adrenal cortex* – stimulates growth and secretion of glucocorticoids |
| prolactin | pituitary (anterior lobe) | hypothalamus – prolactin-releasing factor stimulates | single peptide chain | free in plasma | *mammary gland* – stimulates milk production |
| follicle stimulating hormone (FSH) and luteinizing hormone (LH) *(gonadotropins)* | pituitary (anterior lobe) | hypothalamus – gonadotropin-releasing factors stimulate | glycoproteins – two peptide chains | free in plasma | *testis and ovary* – controls secretion of steroid hormones, sperm production/ovulation and male and female sex characteristics |
| vasopressin (antidiuretic hormone, ADH) | pituitary (posterior lobe) | neurosecretion controlled by electrical activity in hypothalamus, e.g. increased by a rise in the osmotic pressure of blood | small peptide (containing nine amino acids) | free in plasma | *kidney* – water reabsorption *blood vessels* – vasoconstriction |
| cholecystokinin (CCK) | gut (duodenal mucosa) | fatty acids and chyme in digestive system | single chain peptide (33 amino acids) | free in plasma | *gall bladder* – contraction and release of bile *pancreas* – release of digestive enzymes |
| insulin | pancreas (β cells) | increase in blood glucose concentration | two disulphide-linked chains derived from one parent peptide | free in plasma | *liver, muscle and adipose tissue* – promotes glucose uptake and metabolism and general anabolic effects |
| glucagon | pancreas (α cells) | decrease in blood glucose concentration | single peptide chain | free in plasma | *liver* – release of glucose |

(b) Amino-acid-derived hormones

| Hormone | Gland | Major factor(s) regulating hormone secretion | Chemical structure | Mode of transport | Target tissue(s) and some important biological effects |
|---|---|---|---|---|---|
| melatonin | pineal gland | light | derived from tryptophan (hydrophilic) | free in plasma | *brain* – control of seasonal reproductive rhythms |
| adrenalin and noradrenalin (catechol-amines) | adrenal medulla (chromaffin cells) | nervous system | derived from phenylalanine or tyrosine (hydrophilic) | free in plasma | diverse range of effects in *multiple target tissues*: two main receptor systems: α – constriction of blood vessels, mobilization of liver glycogen; β – relaxation of blood vessels, increase in heart rate, broncho-constriction in lungs |
| thyroxin ($T_4$) and triiodothyronine ($T_3$) (thyroid hormones) | thyroid gland | thyroid stimulating hormone (TSH) from anterior pituitary | derived from phenylalanine or tyrosine, iodinated (lipophilic) | >99% bound to circulating plasma proteins | diverse range of effects in *multiple target tissues*: e.g. growth and development; stimulation of metabolic rate and heat production, especially *muscle* and *liver* |

(c) Steroids/cholesterol-derived hormones (lipophilic)

| Hormone | Gland | Major factor(s) regulating hormone secretion | Mode of transport | Target tissue(s) and some important biological effects |
|---|---|---|---|---|
| androgens (e.g. testosterone) | male gonads | luteinizing hormone from anterior pituitary | bound to steroid hormone binding globulin | *testis* – spermatogenesis, maintenance of functional male reproductive system and secondary male sex characteristics |
| oestradiol | ovary | follicle stimulating hormone from anterior pituitary | bound to steroid hormone binding globulin | *uterine endometrium* – cell proliferation |
| progesterone | ovary (corpus luteum) | luteinizing hormone from anterior pituitary | bound to corticosteroid binding globulin | *uterine endometrium* – preparation for egg implantation (pregnancy) |
| glucocorticoids (e.g. cortisol) | adrenal cortex (zona fasciculata) | adrenocorticotropic hormone (ACTH) from anterior pituitary/stress | bound to corticosteroid binding globulin | diverse range of effects in *multiple target tissues*: e.g. control of gene transcription, inhibition of prostaglandin synthesis, inhibition of *immune system* (anti-inflammatory), permissive for the effects of other hormones |
| aldosterone (mineralo-corticoid) | adrenal cortex (zona glomerulosa) | angiotensin II (renin–angiotensin system) | ? | *kidney (distal tubule and collecting duct)* – salt conserving, promoting $Na^+$ uptake, with concomitant $K^+$ loss |

The first group of hormones listed is usually referred to as **peptide** hormones, although their size can vary enormously (from half a dozen amino acids to polypeptides containing hundreds of amino acid residues). Modified amino acids such as adrenalin also fall into this category. Peptide hormones are hydrophilic and so do not easily permeate the cell membrane; they exert their effects by interacting with specific receptors which are expressed on the outer surface of target tissues. You will recall from the previous chapter that these ligand–receptor interactions are able to communicate information into the target cell by a variety of second messenger systems.

By contrast, **steroid** hormones are derived from cholesterol (a component of cell membranes) and are lipophilic. As a consequence, they usually exert their effects via receptors *within* specific target cells, usually in the cytosol or nucleus, since their lipophilic properties enable them to cross the phospholipid bilayer of the cell membrane.

Many peptides are synthesized as larger precursors or **prohormones** which undergo processing to their active form; this occurs either within the gland after translation of the prohormone from the mRNA transcript and prior to release of the hormone, or when the prohormone is bound to its receptor at the target cell. Cleavage enzymes may split the prohormone to yield a number of smaller peptide fragments, many of which may have biological activity. Insulin is an example of a hormone derived from a larger prohormone.

Cells that synthesize peptide hormones or prohormones selectively express genes which encode these proteins and also any enzymes that are involved in processing the final product. Cells that synthesize hormones derived from amino acids or cholesterol are characterized by the presence of unique enzyme systems for hormone synthesis, which are also encoded and regulated by specific genes.

### 4.2.2 Secretion and transport of hormones in the blood

Endocrine glands are highly varied in their location and in the properties of the hormones they secrete, but they have certain features in common, which are summarized in Figure 4.2.

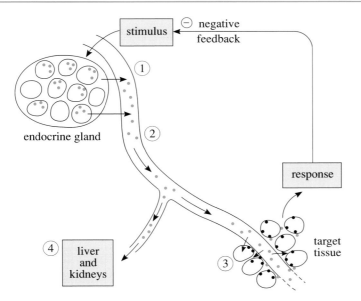

**Figure 4.2** Schematic representation of the generalized communication pathway in the endocrine system. (1) Release of hormone from endocrine gland in response to specific stimuli; (2) transport of hormone in the bloodstream; (3) binding of hormone to specific receptors on target cells, with various biological outcomes; (4) excretion or metabolism of hormone and cessation of its effect on the target cells.

Most endocrine hormones are stored within the cells of the gland in membrane-bound vesicles, prior to release into the extracellular fluid by exocytosis, that is, the fusing of the vesicle with the external membrane, releasing their contents to the exterior (see Figure 3.14 in the previous chapter). The hormone then rapidly diffuses into adjacent capillaries and is swept away in the bloodstream. Hormone secretion is a tightly regulated process controlled by precise and specific stimuli, including electrical signals from neurons, metabolic factors and other hormones. Secretion may be pulsatile, with the frequency of secretory pulses changing in response to a particular external factor, or a rhythmic pattern may be observed, as represented in Figure 4.3. Sometimes this rhythm may be circadian (see Chapter 7).

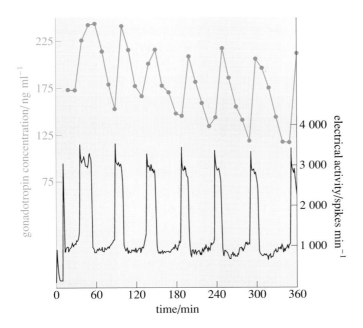

**Figure 4.3** The time courses of plasma gonadotropin concentrations and hypothalamic electrical activity in ovariectomized rhesus monkeys.

▷ Typical concentrations in the bloodstream of circulating hormones are in the range of $10^{-6}$–$10^{-12}$ mol l$^{-1}$. What factors do you think affect hormone concentration? (It will help you to look back at Figure 4.2.)

▶ There are three aspects: (a) the rate of secretion of a hormone from its endocrine gland; (b) the rate of uptake of hormone by its target tissue; and (c) the rate at which the hormone is 'cleared', that is metabolized or excreted from the body.

The rate of hormone clearance is often expressed as a **half-life** or $t_{1/2}$. This term refers to the time taken for the concentration to decrease to half its original value. All hormones are eventually chemically degraded, mainly in the liver; for example, peptides are hydrolysed to smaller fragments, sometimes to their constituent amino acids. Many degradative products are subsequently eliminated by the kidneys.

In most cases it is not so much the absolute amount of circulating hormone, but the *rate of change* in hormone concentration which plays a pivotal role in regulating specific processes in the body. Thus, metabolic and excretory pathways are important in signal termination. Together, the modulation of the rate of secretion and clearance of a hormone ensures that an alteration in physiological circumstances is effectively conveyed (via changes in hormone concentration) to target tissues without too great a delay. However, there will always be some lag in the response while an appropriate adjustment in hormone concentration is made. Endocrine effects at their most rapid occur within a minute; the responses of some hormones develop over a much longer period – days or even months for those involved in growth and development. As we have said, this is one of the important differences between communication via the endocrine and nervous systems.

Many hormones are bound to carrier proteins during their circulation in the bloodstream. Transporting hormones bound to carrier proteins has a number of advantages:

1    The half-life of a hormone is often increased, since the rate of clearance of the bound hormone is reduced compared with that of the 'free' hormone.

2    The solubility of hydrophobic hormones, which are largely insoluble in blood (particularly the steroid hormones such as oestrogen), is increased.

3    The carrier protein acts as a 'reservoir' for the hormone and a buffer which smooths out fluctuations in supplying the hormone to the tissues.

An equilibrium exists between free and bound forms and the free concentration (which is the only form able to interact with a specific receptor) is often quite small (Figure 4.4).

▷ What can you say about the differences in chemical structure between hormone A and hormone B in Figure 4.4?

▶ Hormone A interacts with receptors on the target cell surface, so is likely to be a peptide or other hydrophilic compound. Hormone B is lipophilic since it penetrates the cell membrane.

Not all hormones exhibit wide fluctuations in their circulating concentrations; in some cases, the hormone level is kept constant and in these circumstances transport in the bloodstream via a specific carrier protein is particularly advantageous. Some hormones which have a relatively stable concentration, for example *thyroxin* (see Table 4.1), have a permissive mode of action: this means that only a threshold concentration is necessary to ensure the proper functioning of their target tissues and usually the concentration of hormone in the blood is maintained above this critical level.

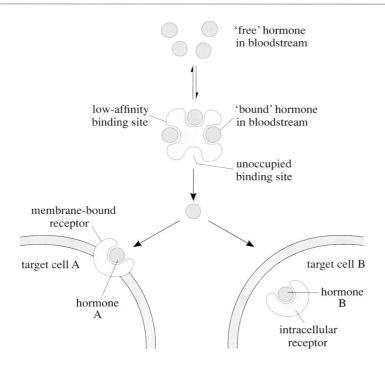

**Figure 4.4** The transport of hormones in the blood as complexes with carrier proteins.

### 4.2.3 Effector actions of endocrine hormones

The target tissue for any hormone is composed of cells that actively transcribe and translate gene(s) that encode receptor proteins which, in turn, bind that hormone. Recall that receptor proteins may be expressed at the cell surface or within the cytosol (as illustrated in Figure 4.4). Once complexed with a receptor, a hormone may exert a variety of effects, including one or more of the categories outlined below:

o   metabolic changes via the activation or inhibition of specific pathways;

o   changes in substrate supply to cells by the regulation of transport systems;

o   alterations in the release of other hormones (or locally acting mediators);

o   alterations in the synthesis or release of other secretory products, such as enzymes or milk;

o   effects on the contractility of smooth muscle, particularly in blood vessels;

o   cell differentiation and proliferation. Hormones that specifically regulate cell division belong to a diverse group of compounds known collectively as *growth factors*. This group of chemical mediators is discussed further later in this chapter (in Section 4.4.4).

Specific examples of these effects can be found by looking back at Table 4.1.

▷   Can you identify endocrine hormones that exert widespread effects on a range of metabolic pathways?

▶   Growth hormone, adrenocorticotropic hormone, insulin, the thyroid hormones and the glucocorticoids all belong to this effector category.

The time taken for these effects to manifest themselves varies considerably. One important factor that determines the rate of response to a hormone (in addition to a change in its concentration) is the nature of the intracellular signal transduction process which the receptor triggers. Usually a peptide or other hydrophilic hormone mediates more rapid responses than a lipophilic mediator since surface membrane-bound receptors are usually coupled to second messengers which have immediate effects within the cell (discussed previously in Chapter 3). Steroids bind to intracellular receptors, which act by regulating gene expression in the nucleus, so there is a latent period between receptor binding and the effect of the hormone on the functioning of the cell.

### 4.2.4 Homeostasis and feedback loops

Many endocrine hormones fulfil the important functions of maintaining homeostasis by controlling the levels of vital physiological parameters within an organism. Most cells require a steady supply of glucose (as an energy source) which is obtained via facilitated diffusion from the extracellular fluid, which itself receives it from the surrounding bloodstream. This means that blood glucose levels need to be maintained relatively constant, normally at around 5 mmol l$^{-1}$. If blood glucose level decreases to less than 3 mmol l$^{-1}$ homeostasis begins to break down, as the brain, which is absolutely dependent on the uptake of glucose from the blood to maintain its functions begins to show signs of failure, and may eventually be irreversibly damaged. However, high concentrations of glucose in the blood can also be harmful; they can lead to irreversible glycation – the addition of glucose molecules to extracellular proteins – which in turn can lead to a loss in function. Excess glucose is also wasted due to excretion in the kidney and water balance is disturbed as a result.

Assuming an adequate food intake, the concentration of blood glucose is maintained within strictly defined limits by the interplay of two hormones, **insulin** and **glucagon**, secreted from cells in the pancreas (β cells for insulin, α cells for glucagon). The output of insulin from the β cells is controlled by negative feedback. A rise in the glucose concentration in the blood stimulates insulin release and this then promotes glucose uptake into target tissues, thus lowering the level of glucose in the blood (see Figure 4.5a). In turn, a fall in the blood glucose level below a set level decreases the release of insulin. This closed loop thus provides one important mechanism to ensure that blood glucose levels do not fall too low. The effectiveness of glucose homeostasis in an individual can be assessed by giving a glucose tolerance test (Figure 4.5b).

**Figure 4.5** The role of insulin in glucose homeostasis.
(a) A schematic diagram of the insulin-dependent mechanisms involved in maintaining blood glucose levels within certain limits.
(b) A glucose tolerance curve showing the effect on blood glucose concentration of a glucose meal in a person who has sustained an overnight fast.

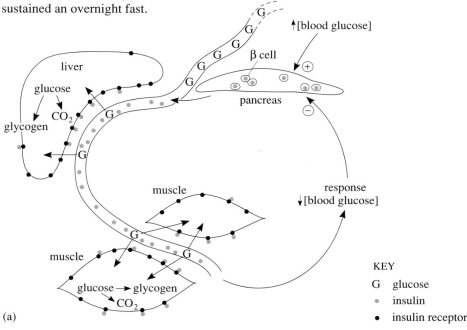

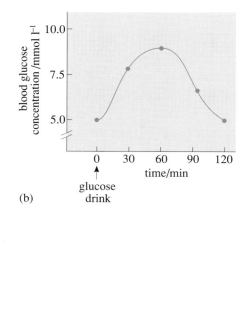

KEY

G   glucose

•   insulin

•   insulin receptor

(a)

▷ What is the other way in which blood glucose levels are prevented from falling too low? (You will find it useful to refer to Table 4.1.)

▶ When the blood glucose concentration is low, glucagon output from the α cells stimulates the release of glucose from the glycogen stores of the liver, thus increasing blood concentration.

Intercellular communication between an endocrine gland and its corresponding target tissue through negative feedback control is an essential feature of hormones which serve homeostatic functions, keeping physiological systems at or near to a set point (as shown in Figure 4.2). Other examples include the control of water balance by the regulation of the pituitary hormone **vasopressin** (also known as antidiuretic hormone, ADH) and the maintenance of adequate thyroid hormone output to replenish $T_3$ and $T_4$ used in metabolism (Figure 4.6).

▷ How many feedback loops are involved in the maintenance of adequate thyroid secretion?

▶ Two. A decrease in the plasma thyroid hormone concentration stimulates the secretion of **thyrotropin releasing factor** (TRF)[*] from the hypothalamus and the release of **thyroid stimulating hormone** (TSH) from the pituitary gland. Together these result in an increase in the stimulation of the thyroid gland, maintaining adequate concentrations in the blood.

▷ The synthesis of thyroid hormones requires iodine, which is incorporated into the hormone molecule. What happens to the concentrations of TRF and TSH in an iodine-deficient individual?

▶ TRF and TSH levels increase, since their release from the hypothalamus and pituitary is stimulated because of low circulating thyroid hormone. However, TSH is unable to increase thyroid hormone release because of the lack of iodine.

For most hormones, the magnitude of a biological effect is proportional to the concentration of hormone present, until a plateau is eventually reached which corresponds to the maximal response (Figure 4.7). The $ED_{50}$ value is the concentration of hormone required to produce a response which is 50% of the maximum.

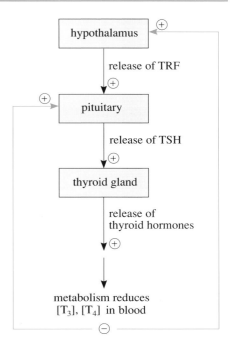

**Figure 4.6** The maintenance of adequate plasma thyroid hormone concentration.

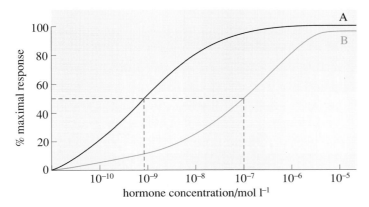

▷ What is the $ED_{50}$ for individual A in Figure 4.7?

▶ Approaching $10^{-9}$ mol $l^{-1}$.

▷ Individual B (in Figure 4.7) shows symptoms of hormone deficiency. What is the $ED_{50}$ value for this patient? How does this account for such clinical observations?

**Figure 4.7** The effect of hormone concentration on the magnitude of the biological response.

[*]Also known as thyroid-stimulating hormone releasing factor.

▶ $10^{-7}$ mol $l^{-1}$. The concentration of hormone required to produce a 50% maximal response is 100 times greater for individual B than individual A. The patient is resistant to the effects of the hormone.

A loss in hormone sensitivity may be due to an alteration in the receptor number and/or signal transduction processes in a target tissue. Maximal responses do not usually require the complete occupancy (i.e. saturation) of receptors, so in most target tissues there are appreciable numbers of spare receptors.

The physiological effect of an endocrine hormone may be modulated in three different ways:

1 The quantity of hormone released in response to a given stimulus may be altered.

2 The rate of clearance of a hormone from the bloodstream may be modulated.

Both these mechanisms contribute to the circulating concentration of the hormone.

3 The sensitivity of the effector cells to a hormone may vary as a result of a change in the receptor population and/or the efficacy of receptor coupling with second messenger systems.

### 4.2.5  Diseases of the endocrine system

Because hormones play so central a part in the body's homeostatic and homeodynamic mechanisms, disorders in hormone production or reception can result in profound metabolic and developmental disturbances. Such disorders are of both physiological and of course clinical interest. Failure of the pituitary to synthesize growth hormone during key developmental periods results in *pituitary dwarfism*, whilst over-production can produce gigantism (*acromegaly*). Failures in thyroxin production in early life either because of a malfunctioning thyroid gland, or a lack of iodine in the diet (recall that the thyroxin molecule contains iodine) can result in irreversible brain damage. Here we will consider just one of the many endocrine dysfunction disorders, the clinical syndrome of **diabetes mellitus.** This is characterized by an elevation of the fasting blood glucose concentration (*hyperglycaemia*) and the excretion of large quantities of glucose in the urine (hence the name *mellitus* for this type of diabetes; the urine tastes sweet; in another, less common type of diabetes large quantities of urine are excreted – *diabetes insipidus*).

It has now become clear that diabetes mellitus itself can result from one of two quite distinct biochemical dysfunctions, only one of which involves an insufficiency of the hormone **insulin**. You will note from Table 4.1 that insulin is important for the maintenance of blood glucose homeostasis. Its rate of secretion from pancreatic $\beta$ cells is enhanced when the blood glucose concentration rises. The increase in circulating insulin stimulates glucose uptake and metabolism in target tissues (mainly muscle and liver, as shown earlier, in Figure 4.5a), in turn resulting in a decrease of blood glucose to its control level of $\sim 5$ mmol $l^{-1}$.

▷ What would be the profile of Figure 4.5b if a glucose tolerance test was performed on a diabetic patient?

▶ Fasting blood glucose levels would be higher, the concentration of glucose in the blood would increase to a much higher level and the time taken for the glucose level to decrease would be longer.

In the first type of diabetes (also termed **Type I**, **insulin dependent** or **juvenile onset**) there is a lack of insulin secretion. In the second form of diabetes (also termed **Type II**, **non-insulin dependent** or **maturity onset**) there is a resistance in target tissues, hence a higher concentration of insulin is required to lower blood glucose levels (as is the case for individual B in Figure 4.7). Type I diabetes arises as the result of the destruction of β cells by the immune system, usually in childhood. In Type II diabetes, some degree of impairment occurs in the structure and distribution of insulin receptors, which interferes with the uptake of glucose; Type II patients are always distinguishable from Type I patients by the presence of insulin (often at higher than normal concentrations) in their blood. The rate of insulin clearance is not adversely affected in either type of diabetes.

## Summary of Section 4.2

One means of long-range communication between the different tissues within an organism is achieved by chemical signalling. Signalling via hormones in the bloodstream is more generally distributed and takes longer (typically minutes or hours) than the much more rapid and specific information transfer of the nervous system.

Hormones are released from endocrine glands; the secretory process is tightly regulated by specific stimuli. Their transport in the blood may be facilitated by binding proteins, and the concentration of circulating hormone is dependent on its rate of secretion and its rate of elimination (i.e. its breakdown and/or removal) from the body. Target tissues are characterized by the presence of specific receptors which are located either inside the cell (if binding hydrophobic ligands) or at the cell surface (if binding hydrophilic ligands). Hormone–receptor interactions instigate appropriate signal transduction pathways inside target tissues and these lead to the final biological response. The sensitivity of such signalling pathways may be modulated providing one means of regulating the effects of the hormone. Often the response acts as a stimulus for controlling the secretion of the hormone from its gland and these feedback loops ensure that both hormone secretion and its biological effects are suitably coordinated. Diseases of the endocrine system may result from a defect at the source of a particular hormone or at an effector tissue. Such hormone disturbances often lead to an impairment in important homeostatic or homeodynamic processes.

## 4.3   Long-range signalling in the immune system

We turn now to another example of a communication network that has to function with a very high degree of precision and accuracy, both to maintain the normal integrity of the body and to ensure a rapid, effective response to infectious organisms. Like the endocrine system, the immune system is a highly dispersed network of cells and macromolecules, which are localized in its own organs (known as *lymphoid organs*) and tissues, but immunologically active cells as well as the molecules they secrete also circulate around the body in the bloodstream. Moreover, the bloodstream is only one of the main carriers of this 'traffic', which is also transported in a parallel system of vessels known as the *lymphatic vessels and capillaries*. Together with the lymphoid organs and tissues, these vessels and their contents constitute the **lymphatic system** (see Figure 4.8 *overleaf*).

Together the vascular and lymphatic systems enable the cells of the immune system to circulate around the body. Cells switch from one transport system to the other by migrating out of blood capillaries as they pass through lymphoid organs

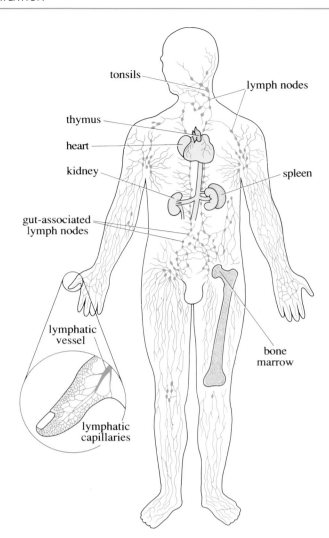

**Figure 4.8** The human lymphatic system consists of the lymphoid organs identified here in blue, connected by an extensive network of lymphatic vessels and capillaries in which a fluid (lymph) transports cells and molecules involved in immune responses. The lymphatic system interconnects with the vascular system (not shown).

(see Figure 4.11 later) and returning to the bloodstream when the lymph flows into it at a point in the neck where vessels from the two systems join. Some of the cells involved in immune responses do not circulate but lead relatively static lives in the lymphoid organs; others are highly motile and 'creep' through the tissues, squeezing between tightly packed cells in every organ in the body. The exception is the brain, where an intact blood–brain barrier prevents the entry of immunologically active cells, unless it has become 'leaky', as is seen in certain disease conditions.

The cells engaged in this perpetual trek around the body are, like red blood cells, derived by cell division from *stem cells* in the bone marrow. To distinguish the two broad 'families' of cells with this shared origin (i.e. red cells and white cells), early microscopists named the cells that were subsequently shown to have immunological functions *white blood cells* or **leucocytes**. As Figure 4.8 illustrates, it is misleading to refer to them as 'blood cells' because many of them spend little or none of their lives in the bloodstream, but the term has stuck. The leucocyte population is extremely diverse and is composed of many different cell types with specialized functions, each with a role in protecting the organism from infection; to describe them all would take us far beyond the scope or aims of this Course. Instead, we have chosen to focus on a single aspect of the immune response in mammals to illustrate intercellular communication across relatively long ranges of time and space. As you study the following account, it is very important that you remember it is a fragment of a complex whole; the mechanism described here would not be adequate on its own to protect you from infection.

### 4.3.1 The antibody response to infection

In Book 1 you learned about the molecular structure of immunoglobulins, commonly known as **antibodies**. We can now examine their role in the immune response to infection and the signals that trigger their synthesis. In order to focus on cell-to-cell communication, rather than discuss the whole of immunology, we have had to omit many interesting details from the events that occur between the moment that an infectious organism is detected in the body and the arrival, some time later, of the first antibodies that bind to the organism.

Antibodies come in different shapes and sizes, but all are built on the same basic plan (described in Book 1, Chapter 3; you will find it helpful to look back at Figures 3.29 to 3.31). Remember that the Y-shaped antibody protein consists of four covalently bound polypeptide chains: two identical heavy chains and two identical light chains. The conformation of these chains in space produces two identical 'clefts', each of which can bind to a ligand with the corresponding shape and charge profile. These receptor surfaces are usually referred to as **antigen binding sites** and any cell or macromolecule to which an antibody can bind is termed an **antigen**. The antigen binding site is quite small: it can only accommodate a ligand composed of between 5 and 15 amino acid residues, or sugar units, or a combination of both. Less commonly, nucleic acid residues, either alone or complexed with protein, may be bound by antibodies. Any cluster of residues constituting a ligand that can be bound by an antibody molecule, is known as an **epitope**.

The range of unique epitopes that exists on the surface, or within the structures, of infectious organisms such as bacterial cells or viruses, is very large: mathematical modelling predicts that the immune system of mammals can detect at least $10^7$ different epitopes and perhaps as many as $10^8$. Furthermore, infectious organisms are complex antigens; each species or strain has a unique assortment of thousands of different and overlapping epitopes on its surface (Figure 4.9a). This diversity has created the evolutionary pressure towards an ever-more complex immune system, capable of transmitting detailed information between its members about the presence of infectious organisms. When such an organism gets into the body, it sets off a network of intercellular signals, which have a variety of outcomes. One of the most important is that they initiate the synthesis of an array of antibodies, each with its own *specificity,* distinguished by differences in the conformation of their antigen binding sites. Each antibody 'species' (as immunologists call them) will only bind to a particular epitope on the infectious organism, and a range of

**Figure 4.9** Schematic diagram of the outer membrane of a hypothetical infectious cell. (a) Surface view showing the areas of many different epitopes, some of which overlap; each unique epitope has a characteristic shape and charge profile. (b) Sectional view showing each unique epitope bound by a molecule of antibody with a binding site (receptor surface) that is specific for that epitope (ligand).

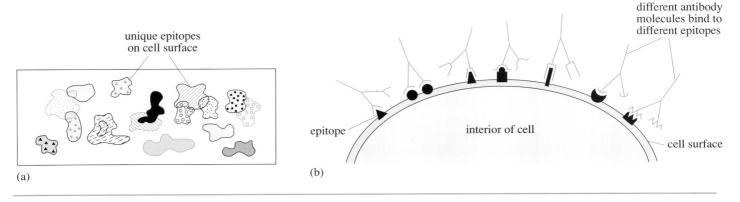

unique epitopes on cell surface

different antibody molecules bind to different epitopes

epitope

interior of cell

cell surface

(a)

(b)

affinities and concentrations exists within a given pool of antibodies. This results in the huge diversity of antibody responses that can be induced to different epitopes found on the same antigen and, as a consequence, maximizes the number of antibody molecules that can bind to an infectious organism (Figure 4.9b).

There are a number of reasons why an antigen such as a bacterial cell is destroyed more quickly if it has antibodies bound to its surface than if it has not. The antibodies themselves have no effect on the antigen, but they 'label' it for destruction by other mechanisms in the immune system. For example, several sorts of leucocytes are *phagocytic*, that is they engulf antigens by endocytosis and destroy the contents of the resulting endosome by fusing it with lysosomes which, as you will remember, contain a potent cocktail of degradative enzymes and oxidants. These phagocytic cells have receptors on their surfaces which bind to an invariant ligand in a constant region of the antibody molecule; this facilitates endocytosis by improving the 'grip' of the phagocytic cell on its target (Figure 4.10). Antibodies bound to antigens have many other effects on aspects of the immune system, which do not concern us here; simply note that they all increase the rate of elimination of the antigen.

**Figure 4.10** Antibody molecules bound to an antigen (e.g. a bacterial cell as shown here) facilitate its destruction by phagocytic cells, which have surface receptors for invariant residues in the antibody structure. (*Note:* Nothing in this diagram has been drawn to scale!)

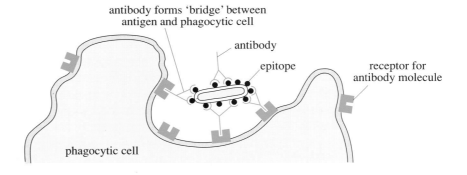

▷ In Figure 4.10, are the antibody molecules behaving as receptors or as ligands?

▶ Both: the antibody contains binding sites for ligands (epitopes) on the antigen, while simultaneously another part of the antibody molecule is acting as a ligand itself, to which a receptor on the phagocytic cell can bind.

You might have imagined that antibodies would be synthesized and secreted by cells in the immediate vicinity of the point at which the infection entered the body. However, this is not always true. In most infections, the main site of antibody synthesis is in the nearest lymphoid organ, which may be several centimetres away (look back at Figure 4.8). The news that an infection has got into the body has to be 'translated' into a readable form and relayed some distance to a number of different cell types involved in the production of the appropriate antibodies. We will look at the two major cell types in a moment: they are called *B cells* and *helper T cells*. Their interaction, in a precise sequence of communication events, ensures that only the correct species of antibodies are produced (i.e. those with binding sites for epitopes on the infectious organism) and that antibody secretion ceases when the infection has been eliminated.

▷ What are the advantages for the 'host' organism (in which the immune response is occurring) of restricting antibody production in this way?

▶ Synthesizing excess antibodies, or those with binding sites for epitopes *not* found on the current source of infection, wastes biochemical resources, but more importantly it exposes the host animal to possible 'bystander' damage if antibodies bind inappropriately to the host's own cells.

As a consequence of the specificity with which cells of the immune system recognize and communicate with each other, targeting of an immune response is very precise. Receptors on the surface of one cell will bind either to ligands on the surface of another, or to signalling molecules secreted by a collaborating cell. The response proceeds a step at a time and will only progress towards antibody synthesis if the correct sequence of signals is received. Thus, 'failsafe' mechanisms are built into the system and the response can be rapidly scaled down if essential conditions are not met.

One of the events involved in initiating antibody synthesis occurs when epitopes on an invading antigen are bound by antibody molecules embedded in the surface membrane of a class of leucocyte called **B cells** (or B lymphocytes, to give them their full title). Mature B cells each have about $10^5$ molecules of antibody inserted in their cell membrane, with the antigen binding sites facing outwards into the extracellular fluid (Figure 3.29 in Book 1 shows this schematically). The antibodies on each individual B cell have identical binding sites, so each B cell can only bind to an extremely restricted range of epitopes with conformations that 'fit' its specific antigen binding sites. In effect, each B cell is restricted to 'recognize' a single target epitope and is unable to interact with any other epitopes because they cannot be bound by that B cell's surface antibodies. Complete coverage of all the possible epitopes that might be encountered in life is achieved because the population of B cells is highly varied with respect to the range of antigen binding sites available; the population contains at least a few individuals with surface antibodies capable of binding one of at least $10^7$ different epitope shapes. The ability to generate such a huge diversity of antibody specificities within the B cell population of a single organism rests on the rearrangement (or somatic recombination) of the genes encoding different sections of the antibody molecules; in effect, a relatively small number of genes are 'shuffled' into a huge variety of different sequences, which in turn generate the diversity seen in their protein products – the antibodies. Somatic recombination as the basis of antibody diversity is discussed in detail in Book 4, Chapter 7.

Binding of the epitope to the surface antibodies of a B cell generates signals within the cell, via the signal transduction mechanisms described in Chapter 3. These signals are one of the mechanisms that induce B cells in an infected area to express increased numbers of *homing receptors* and **adhesion molecules** on their surface. The B cells are swept along in the circulation until they enter the nearest lymphoid organ (usually a lymph node, see Figure 4.11 *overleaf*), where they are trapped when the homing receptors and adhesion molecules on the cell surface bind to ligands on the surface of cells lining the entry to the organ. (Some of these ligands have been quaintly named *addressins* because they signal to the B cells that they have reached the correct 'address'.) Later in this chapter, we will look at adhesion molecules in a little more detail (Section 4.5.1). Alternatively, the antigen may be swept along in the circulation until it enters a lymphoid organ and encounters resident B cells with receptors to which it can bind.

For most B cells an additional signal is required to trigger the next stage in the process that leads ultimately to the synthesis and secretion of antibodies into the lymphatic and vascular circulations. Most B cells that have already bound to an epitope still require an activating signal from another type of lymphocyte called

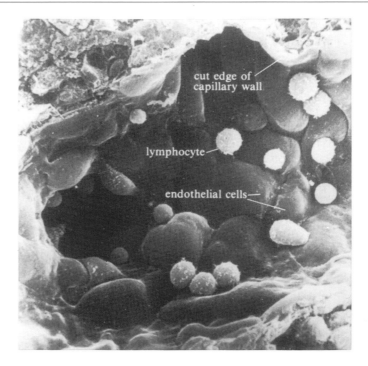

cut edge of capillary wall

lymphocyte

endothelial cells

**Figure 4.11** Scanning electron micrograph of lymphocytes (B cells or T cells) migrating into a lymph node by squeezing between the endothelial cells in the walls of blood capillaries passing through the node. Receptor–ligand interactions between proteins in the surface membranes of the lymphocytes and the endothelial cells 'trap' lymphocytes which have already been in contact with an appropriate antigen.

a **helper T cell**. The signalling molecules involved are members of a diverse group of low molecular-mass proteins known collectively as **cytokines**, which are synthesized by cells involved in an immune response and in turn stimulate or inhibit the activity and differentiation of other leucocytes in the vicinity. We will say a bit more about cytokines later (Section 4.4.1), under the general heading of 'short-range signalling mechanisms'. Cytokines operate across extremely small gaps between the helper T cell and the B cell, created when the cells are briefly held together by adhesion molecules on the surface of one cell binding to appropriate receptors on the surface of the other. During this brief moment of cell-to-cell contact, cytokines in the communicating cells become localized close to the surface membrane in the contact zone, before being released into the gap.

The B cell has surface receptors for particular cytokines which are secreted into the contact zone by a collaborating helper T cell. These cytokines must bind to the B cell before it can replicate and differentiate into a clone of **plasma cells**, which are in effect 'antibody-factories'. The plasma cells derived from a single B cell are all identical and produce identical antibodies, all of which have antigen binding sites that bind only to the corresponding epitopes on the original antigen. The antigen is thus responsible for selectively activating only those B cells from the population that have surface antibodies which bind to its epitopes. Providing the necessary cytokines are delivered to these B cells by the helper T cells, they transform into plasma cells which synthesize large quantities of the *same* antibodies the 'parent' B cell carried on its surface. These antibodies are secreted into the lymph and bloodstream as they pass through the lymphoid organ and so circulate around the body, binding to the antigen wherever it is encountered.

▷ What is the principal advantage of this long-range communication network?

▶ Antigens that enter the body at a single site trigger an antibody response in the nearest lymphoid organ, which subsequently becomes systemic, i.e. distributed throughout the body, giving widespread protection against additional entries of the same infection and against the dispersal of that infection around the body in the bloodstream.

This sequence takes time: if the original antigen has never before been encountered by the immune system, it takes several days before a large enough clone of plasma cells has developed to produce effective amounts of the correct antibodies in the circulation. In this time lag, symptoms of the infectious disease usually appear. However, some members of the expanded clone differentiate not into plasma cells but into long-lived *memory cells*; if the antigen is encountered for a second or subsequent time, the memory cells form the fast-reacting focus for an accelerated immune response which may be sufficiently effective to eliminate the infection before symptoms arise.

In describing the sequence of events leading to antibody synthesis, we have intentionally left out many details, but an important aspect of the communication between B cells and T cells has to be introduced. We have already mentioned that the helper T cells are the source of the essential cytokines required for complete B cell activation to many (though not all) antigens. However, the T cells themselves require a permissive signal from the B cells before they will secrete the essential cytokines. This circularity and reciprocity in communication between cells is typical of the regulatory mechanisms within the immune system: activation of one cell type requires a signal from another, which in turn requires a signal from the first cell, and so on until a number of communication steps have been completed in a precise sequence. A feature of this sequence is that each chemical signal from one cell to the other triggers an increase in either the *density* or the *affinity* of surface molecules on the opposing cell; these molecules are either receptors or ligands involved in cell–cell communication or they are adhesion molecules holding the two cells together.

▷   What effect will these changes in density or affinity at the cell surface have on communication between the two cells?

▶   It will facilitate communication by stabilizing their physical contact, by increasing the density of ligands and receptors (some of which were previously at too low a density to participate in effective signalling), and by increasing the affinity of receptors for ligands (some of which previously had too low an affinity to bind the ligand). The overall effect is to lower the 'communication threshold' between the two cells, enabling them to transmit and coordinate effective signals to each other.

A closer look at the interaction between B cells and helper T cells also reveals that they 'recognize' and respond to the same antigen in quite different ways. As you already know, B cells bind to epitopes on the original unprocessed antigen, using surface antibodies as their antigen receptor. Having 'captured' an antigen by this binding event, B cells then internalize it and process it into a form that helper T cells can recognize. Unlike the B cells, helper T cells cannot bind to the intact antigen, but bind only to peptide fragments of the antigen which are 'presented' on the surface of so-called **antigen-presenting cells** (or APCs). Several different cell types in addition to B cells can function as APCs, but we will focus on B cells as presenters of processed antigen here.

APCs have in common the ability to phagocytose antigens and break them down into peptide fragments, which are then transported back to the cell surface and 'displayed' in the cell membrane. In this location, the peptides derived from the antigen are already associated with naturally occurring proteins – the class II *major histocompatibility complex* molecules (abbreviated to MHC II, see Book 1, Chapter 3) – in the surface membrane of the APC. Each MHC II molecule has a central cleft in its structure, in which a peptide fragment of the antigen sits (the arrangement has been described as looking rather like a hot dog in a bun). The association of peptide fragments of the antigen with the MHC molecules in this precise structural relationship creates a novel molecular complex on the surface of the APC, which it then 'presents' to helper T cells (Figure 4.12). Regions of the surface of this molecular complex form a unique ligand, constructed partly from sequences in the antigenic peptide and partly from sequences in the MHC molecule. Only those helper T cells with the correct receptor for this ligand can bind to it. The antigen receptors of a helper T cell can only 'recognize' antigen and bind to it in this 'processed' form, i.e. as a peptide fragment held in the cleft of a MHC II molecule on the surface of an antigen-presenting cell.

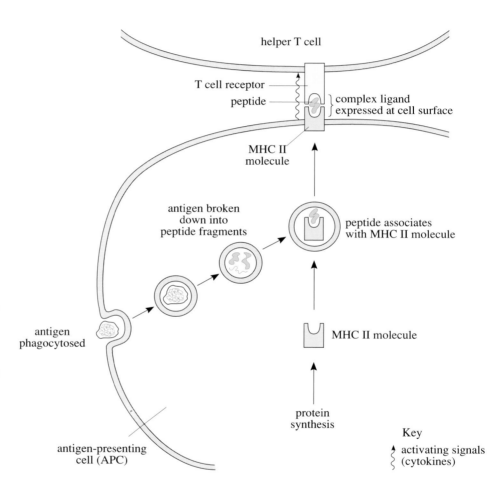

**Figure 4.12** Antigen presentation to helper T cells involves the expression of a complex ligand formed from a peptide fragment of the antigen held in a precise spatial relationship in the structure of a naturally occurring MHC II molecule on the surface of the presenting cell. The receptor for antigen on the helper T cell binds with both components of this ligand. B cells are one of the major cell types capable of presenting antigen to helper T cells.

We can now summarize the events leading to the synthesis and release of antibodies as part of an immune response to an antigen (Figure 4.13). The epitopes on the surface of the 'native' unprocessed antigen are bound by surface antibodies on resting B cells; this is an obligatory first step in B-cell activation. The B cell engulfs the antigen and processes it into peptide fragments, which associate with MHC II molecules inside the cell. This molecular complex is transported to the B cell surface where it is expressed externally and 'presented' to helper T cells. Binding of the T cell's antigen receptor to the peptide-MHC ligand on the B cell's surface initiates a series of communication signals back and forth between the two cells. These chemical signals induce an increase in the density and affinity of surface molecules engaged in cell–cell communication and adhesion, 'uprating' the effectiveness of signalling between the two cells. The end result is the release of certain cytokines by the helper T cell which complete the activation of the antigen-primed B cell. The B cell differentiates into a clone of plasma cells, which synthesize and secrete antibodies.

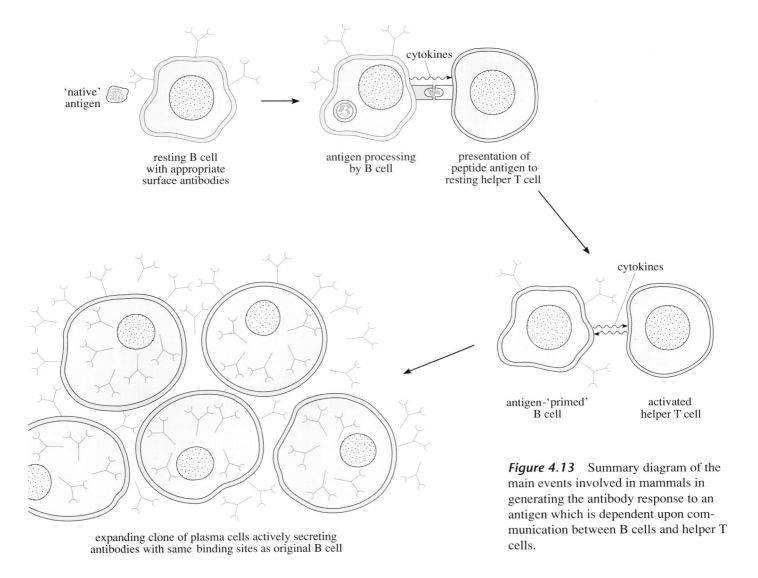

'native' antigen

resting B cell with appropriate surface antibodies

antigen processing by B cell

cytokines

presentation of peptide antigen to resting helper T cell

cytokines

antigen-'primed' B cell

activated helper T cell

expanding clone of plasma cells actively secreting antibodies with same binding sites as original B cell

**Figure 4.13** Summary diagram of the main events involved in mammals in generating the antibody response to an antigen which is dependent upon communication between B cells and helper T cells.

In the next section we discuss short-range signalling mechanisms in more detail, including some of those touched on above, but before we do so, it is important to remind you that the antibody response we have just described is but one aspect of the immune response to an infection. Full protection against the range of infectious organisms and their toxins requires a great deal more than antibodies, but this is beyond the scope of this Course.

## Summary of Section 4.3

The responses of the immune system to infectious organisms or other antigens rely on a network of long-range and short-range intercellular signals. Immunologically active cells continually circulate around the body in the vascular and lymphatic systems and migrate through the tissues. Contact with infectious organisms or their products induces a coordinated immune response, which relies on close contact and direct communication between cells. Cell-to-cell contact is facilitated by a number of different adhesion molecules (described in more detail in Section 4.5.1 of this chapter). Adhesion molecules are glycoproteins expressed at the surface membrane of all cells (see also Book 1, Chapter 1); they are particularly important in trapping leucocytes (such as B cells and helper T cells) in lymphoid organs, once they have bound to antigens. Cytokines are short-range signalling peptides (described in more detail in Section 4.4.1 below), which are synthesized by cells involved in coordinating and targeting an immune response. For example, certain cytokines secreted by antigen-presenting cells (APCs) activate the helper T cells to which they are also presenting peptide fragments of antigen. In turn, the helper T cells secrete cytokines which promote the replication and differentiation of B cells which have previously bound to unprocessed antigen via antibody molecules in its surface membrane. B cell replication generates a clone of plasma cells, which synthesize and secrete antibodies with binding sites that recognize epitopes on the original antigen. This facilitates the elimination of the antigen from the body, for example by increasing the efficiency of phagocytosis. Antibody synthesis is only one aspect of the immune response to infection, but it demonstrates the complexity and diversity of the immune system, which has evolved as a consequence of selection pressure from the huge diversity of infectious organisms in the environment.

## 4.4  Short-range signalling mechanisms

The release of short range chemical mediators – more usually described as **local hormones** – provides an important means by which individual cells can communicate with others nearby. This route bypasses the need for physical cell-to-cell contact, but the effects are only manifested within a very narrow radius of their site of release since these compounds are rapidly degraded. Local hormones may exert effects on the cells from which they originated when they are said to behave in an **autocrine** manner. As well, or alternatively, they may act on adjacent or nearby cells, in which case they are described as **paracrine** hormones. As you will learn from some examples discussed later, many chemical compounds fulfil roles as locally acting hormones.

▷ What factors might affect the range of action of a local hormone?

▶ (a) The quantity of the mediator released.

(b) The rate of diffusion of a mediator between cells (in the extracellular space and sometimes through phospholipid bilayers).

(c) The chemical stability of the hormone in the extracellular fluid (i.e. its tendency to decompose spontaneously).

(d) The presence of enzymes or other factors which catalyse the inactivation of the hormone in the extracellular fluid.

In addition to a shorter half-life and rapid turnover, there are other distinct differences between long- and short-range intercellular communication. First, the majority of local mediators are not stored pre-formed within cells, but are instead synthesized in response to a specific stimulus. Second, lipophilic paracrine/autocrine factors simply traverse the cell membrane by diffusion once they have been synthesized, although peptide factors do require a specific mechanism of release, namely exocytosis. Thus, in contrast to endocrine compounds, release is not a major event which regulates paracrine activity: instead it is the process of synthesis which is an important control point for the action of local hormones.

Finally, the synthesis and release of local hormones are not restricted to a specific endocrine gland, but occur in nearly all tissues. Their actions rely on the presence of specific receptors, but again these are widely distributed rather than located in discrete target cells. Since the ability to secrete and respond to a diverse range of local mediators is ubiquitous, discrimination is achieved by the regulation of mediator synthesis by specific stimuli, the affinity of the receptors for ligands and the concentration gradient which is generated by the diffusion of a paracrine agent from its site of production. Hence, despite the widespread existence of multiple paracrine and autocrine hormone systems within an organism, their actions remain specific and local. Examples of some physiologically important local mediators are discussed in further detail below. In addition, the example of the antibody response illustrates that the immune system uses many different short-range signalling mechanisms, some of which hold cells together and others that transmit chemical messages across small but variable gaps as cells in a fluid matrix approach each other. The short-range signalling mechanisms described in this section, however, are not unique to the immune system; several of them are employed by many different cell types in the body. At the end of this chapter we will discuss chemical signalling across fixed 'gaps' between cells in a solid tissue matrix.

### 4.4.1 Cytokines

Cytokines are a large family of locally acting peptide mediators which are synthesized and released by a spectrum of cells whose function involves the coordination of the immune response and the growth and differentiation from bone marrow stem cells of all the different kinds of leucocytes (white cells in the bloodstream and lymphatic system). In contrast to many local hormones, they may be stored pre-formed in intracellular vesicles before release. Their synthesis may be triggered by other cytokines (as part of a cascade) or by interaction with native or processed antigens, as described in Section 4.3.1. Cytokines are active at extremely low concentrations ($10^{-15}$ mol l$^{-1}$), and generally have multiple effects on the growth and differentiation of the cells to which they bind. It follows that for every cytokine, there is a specific receptor on the target cell.

Important cytokines include at least 13 different *interleukins,* which have major roles in cell-to-cell communication in the immune system, principally the activation of the cell to which the interleukin binds (as shown earlier in Figure 4.13). The other major classes of cytokines are the *colony stimulating* and *cell growth factors,* which promote cell proliferation in specific target populations of leucocytes, thereby 'scaling up' an immune response; two different *tumour necrosis factors,* which are involved in inflammatory reactions around sites of infection and tumour growth (these factors were originally identified as substances having some action against tumours, hence their name, but they have a far wider functional relevance than this implies); and the *interferons,* which (among other effects) 'interfere' with virus replication and stimulate the expression of important cell-surface molecules involved in cell–cell interactions in an immune response. Distinguishing between the different cytokines has been extremely difficult, both biochemically and functionally, but the wide range and importance of their effects make them the subject of much current research.

### 4.4.2 Nitric oxide

Following its identification as a vasodilatory mediator (that is, one which increases blood flow by relaxing the blood vessels) in 1987, interest in **nitric**

**oxide** (NO), a gaseous inorganic molecule (mentioned briefly in the previous chapter), has revealed that NO (which is not to be confused with the anaesthetic nitrous oxide or laughing gas, $N_2O$), performs a number of crucial functions in mammals as a local messenger molecule. It may be that NO is but one of a whole range of such gaseous messengers. The gas ethylene has long been known to play an important signalling role in the growth and development of plants. There have also been suggestions, unconfirmed at the time of writing this Course, that the gas carbon monoxide, CO, better known as a poison as a consequence of its capacity to bind to and inactivate haemoglobin, may also serve as such a local messenger.

The instability of nitric oxide is apparent from its chemical structure, $^\bullet N{=}O$. It is a *free radical*, possessing an extra electron which renders the molecule highly reactive and short lived. Its diverse biological properties are not shared by any of its breakdown products, and its non-polar structure and small size allow it to diffuse readily through the extracellular space, cross the external cell membrane and serve as a signal for an intracellular cascade. Because it can diffuse rapidly, once released at one site it can quickly reach many nearby cells to engage them in coordinated activity before being destroyed. Hence, nitric oxide meets the credentials of a local hormone.

Nitric oxide synthesis is catalysed by the enzyme *nitric oxide synthetase* (NOS), in a two-stage process by the oxidation of a nitrogen atom present on the amino acid arginine; citrulline is formed as a by-product (Figure 4.14). The activity of NOS is rate-limiting for NO synthesis and is itself responsive to a variety of stimuli. There are two isoenzymes of NOS, and each is regulated in a different way. One form of NOS is constitutively expressed, that is, it is continually produced in cells, particularly neurons and endothelial cells, and is activated by calcium. The other is calcium insensitive, and inducible by chemical stimuli, such as cytokines.

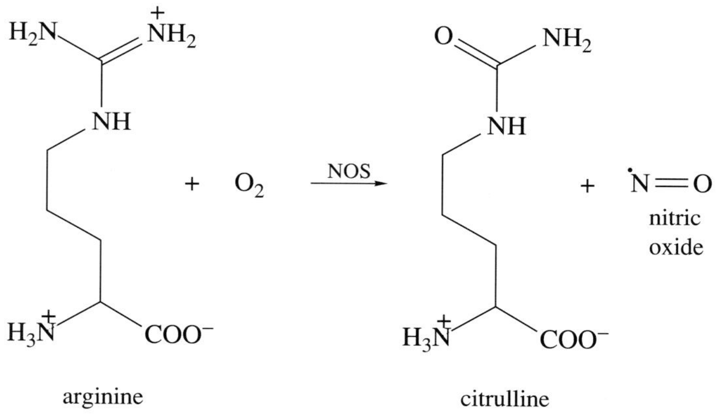

**Figure 4.14** The biosynthesis of nitric oxide.

Measurements of the rate of the spontaneous decomposition of nitric oxide in the presence of oxygen indicate that the molecule has a half-life of less than 30 seconds. However, since nitric oxide diffuses rapidly it is able to interact with receptors in adjacent cells despite its marked instability. The most important effector target of nitric oxide is a rather unusual receptor – the enzyme guanylyl cyclase (mentioned in Chapter 3). Recall that the soluble, cytosolic form of guanylyl cyclase is activated by direct binding of NO, which chemically reacts with the enzyme's haem prosthetic group. The resulting increase in intracellular cyclic guanosine monophosphate concentration has a number of consequences, most significantly the relaxation of smooth muscle: this is the mechanism of nitric oxide's vasodilatory action.

The endothelial cells which line the walls of blood capillaries are the major source of the nitric oxide, which then exerts its effects on the smooth muscle cells beneath. Stretch-activated calcium channels which are present in endothelial cells open transiently as a result of the distension created by the pulsatile flow of blood created by the heartbeat and this is a trigger for the production of nitric oxide. In addition, agonists such as the neurotransmitter acetylcholine (produced in response to nervous stimulation) cause vasodilation and this is also mediated by a calcium-dependent activation of NOS in endothelial cells (Figure 4.15).

The physiological importance of nitric oxide in the control of blood flow has been demonstrated by the use of specific inhibitors.

▷ The arginine analogue, *monomethylarginine*, blocks the synthesis of NO in cells which express NOS. How does it achieve this block? (The name of the compound should give you a clue.)

▶ Since the compound is structurally similar to arginine, it competes for binding at the active site of NOS, but is unable to undergo chemical transformation to NO. Thus the conversion of arginine to NO is inhibited.

The administration of monomethylarginine to experimental animals causes *vasoconstriction* and a significant rise in blood pressure. This suggests that nitric oxide is continually generated in the vascular endothelium and is responsible, at least in part, for the maintenance of basal vasodilator tone.

Nervous tissue is another major source of nitric oxide: NOS is particularly abundant in cerebral tissue, and often localized at nerve terminals. Its possible role in synaptic signalling will be considered in Section 4.5.4 below. Neurons which terminate in vascular beds are, via the release of NO, able to mediate vasodilation by a direct, endothelium-independent relaxation of smooth muscle: this process has been shown to be important in controlling blood flow to the penis during erection and a lack of NO release has been implicated in impotence.

However, NO is not always employed as a transducer molecule. In the immune system, its free radical structure is exploited for the direct attack of pathogens. Specific immune cells called **macrophages** are a major source of NO during infection. The molecule is synthesized within these cells following the cytokine-mediated induction of NOS and its release contributes to their bactericidal activity. In severe infection, the immune response causes blood cytokine levels to increase significantly. In turn, large amounts of nitric oxide are generated in endothelial cells and activated macrophages resulting in a dramatic vasodilatory response. This may lead to cardiovascular failure and circulatory shock.

### 4.4.3 Eicosanoids

This is a general term for a large family of local hormones (of which more are steadily being discovered) which are derived from 20-carbon polyunsaturated fatty acids. *Arachidonic acid* (which contains four double bonds, C20 : 4) is the most common lipid precursor of these compounds (Figure 4.16a).

Eicosanoids share a number of functional characteristics with nitric oxide, despite their very different chemistry. They are widely distributed throughout animal tissues, where they exert a diverse range of localized physiological actions (see Table 4.2 later). They have short half-lives (typically less than a minute), in parallel with all paracrine/autocrine factors, so unless abnormally high amounts

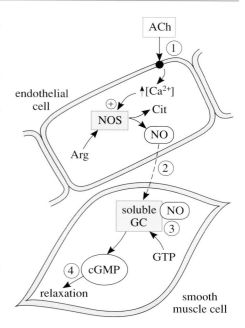

**Figure 4.15** Endothelium-dependent vasodilation mediated by the release of nitric oxide. (1) Acetylcholine (ACh) activates nitric oxide synthetase (NOS) by an increase in intracellular calcium. (2) Nitric oxide (NO) is synthesized in endothelial cells and diffuses into smooth muscle cells beneath. (3) NO binds to and activates guanylyl cyclase (GC), stimulating an increase in cyclic guanosine monophosphate (cGMP). (4) cGMP mediates smooth muscle relaxation – vasodilation. Other abbreviations: Arg, arginine; Cit, citrulline.

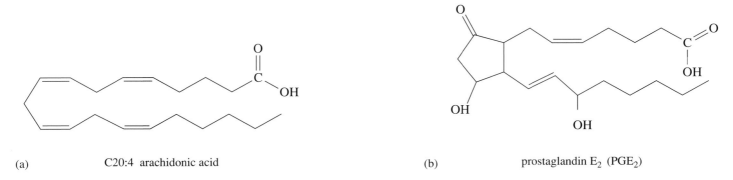

(a)  C20:4 arachidonic acid

(b)  prostaglandin E$_2$ (PGE$_2$)

**Figure 4.16** Structures of (a) C20 : 4 arachidonic acid and (b) prostaglandin E$_2$ (PGE$_2$).

are synthesized, their effects are restricted to cells that are located in the vicinity of their source. Important members of this family include the **prostaglandins**, **thromboxanes** and **leukotrienes**.

Eicosanoids readily permeate lipid bilayers, so they diffuse from their site of synthesis into surrounding cells without a need for a specific release mechanism. The concentrations of these local hormones are regulated by the activity of their biosynthetic pathways. However, the rate of eicosanoid production is largely determined by the concentration of the pathway precursor, typically arachidonic acid, so that an increase in its availability stimulates eicosanoid synthesis as a consequence of a mass-action effect.

▷   How does this process contrast with the regulation of nitric oxide production?

▶   In this case, the pathway is not regulated by the availability of the precursor (arginine), but instead is controlled by the activity of the enzyme nitric oxide synthetase.

Thus, eicosanoid production is affected by any process which regulates arachidonic acid release. The cell membrane is a rich source of such polyunsaturated fatty acids; fatty acids may be liberated from the phospholipid bilayer (where, as you will recall from Book 1, Chapter 1, they occur esterified to a glycerol backbone) by the action of the enzyme phospholipase A$_2$ (Figure 4.17). This enzyme is itself stimulated by an elevation of intracellular calcium concentration and many receptor transduction systems enhance its activity in this way (for example, the peptide hormone vasopressin, which regulates water uptake – look back at Table 4.1). Intracellular phospholipase A$_2$ activity may also be inhibited by *lipocortins*. These are substances that are synthesized in cells when they are exposed to steroid hormones such as glucocorticoid.

▷   What would be the effect of vasopressin on prostaglandin synthesis within the kidney collecting duct if blood glucocorticoid levels are also increased? (Figure 4.17 should help.)

▶   Vasopressin stimulates phospholipase A$_2$ activity, while the expression of lipocortins as a result of the glucocorticoid inhibits the liberation of arachidonic acid and other fatty acids from membrane phospholipid. Hence the production of prostaglandins in response to vasopressin is blocked.

Most cells contain specific receptors for different types of eicosanoids. These are usually found at the cell membrane (rather surprising considering the hydrophobic nature of these ligands) and the second messenger systems of some have been identified, including G proteins which may stimulate (as well as inhibit) adenylyl

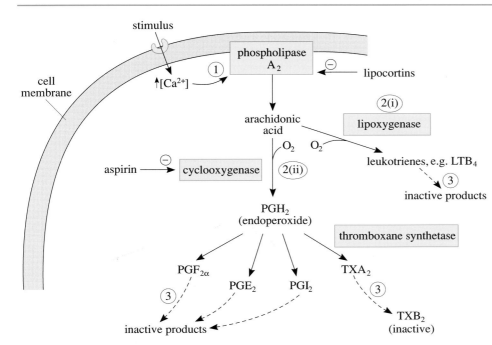

**Figure 4.17** The pathway of eicosanoid synthesis. (1) The mobilization of arachidonic acid from membrane phospholipid by the activation of phospholipase $A_2$. (2) Metabolism of arachidonic acid by (i) lipoxygenase pathway, producing leukotrienes and (ii) cyclooxygenase pathway, resulting in the formation of prostaglandins and thromboxanes. (3) Inactivation of eicosanoids by spontaneous or enzyme-catalysed decomposition. Note how agents such as aspirin can affect these steps.

cyclase and G proteins which activate the phosphatidyl inositol system. Guanylyl cyclase activity and ion channels have also been shown to be coupled to eicosanoid receptors.

Endocrine and paracrine systems are often highly interactive. Indeed, many local hormones – eicosanoids in particular – function as modulators of endocrine hormone action, rather than having direct effects within a tissue. This is evident if we further consider the example discussed above. In kidney collecting duct epithelium vasopressin activates adenylyl cyclase (in addition to phospholipase $A_2$), resulting in an increase in the concentration of cAMP, as well as stimulating the formation of prostaglandin $E_2$ (PGE$_2$, structure shown earlier, in Figure 4.16b). An elevation in cAMP concentration in these cells mediates the reabsorption of water – this is the major physiological effect of vasopressin. However, the simultaneous production of PGE$_2$ by vasopressin decreases the direct effect of the hormone because the prostaglandin receptors found in this tissue are coupled (via $G_i$ – inhibitory G protein, see Chapter 3) to the inhibition of adenylyl cyclase, so in this example the PGE$_2$ is acting through an autocrine loop. The overall effect of vasopressin on water retention within an organism is thus determined by cAMP, and the balance between the direct activation of adenylyl cyclase (via the vasopressin receptor) and the indirect prostaglandin-mediated inhibition determines the intracellular concentration of this important second messenger.

As shown in Figure 4.17, a range of eicosanoid products may be formed following the release of arachidonic acid from membrane phospholipid. Oxygenation via a **cyclooxygenase** enzyme yields the endoperoxide PGH$_2$, and this serves as the precursor for bioactive prostaglandin and thromboxane molecules. The types of prostaglandins or thromboxanes that are produced depend on the enzymes which are found within a particular cell, for example PGEs are the predominant species in the kidney collecting duct, while thromboxanes (TXAs) are the major products in blood platelets where they are responsible for promoting aggregation.

Prostaglandins of different types (e.g. PGE, PGD, PGF, etc.) have varying effects and potencies in different tissues due to the presence of distinct receptors with different binding affinities. All are rapidly inactivated, either by spontaneous decomposition (e.g. $TXA_2$ is unstable and breaks down to the stable, biologically inert $TXB_2$) or by enzymically catalysed reactions.

The *lipoxygenase* pathway leads to the formation of another class of eicosanoids, *leukotrienes*, first identified as the secretory products of immune cells. They also cause broncho-constriction in the lung: leukotriene hypersecretion has been implicated in the development of asthma.

Eicosanoid synthesis may be inhibited pharmacologically. The most well-established inhibitor is *aspirin* (acetyl salicylate), which specifically blocks the action of cyclooxygenase by irreversible acetylation, therefore inhibiting the formation of prostaglandins and thromboxanes. Many of the effects of aspirin can be explained by its inhibitory effect on the synthesis of eicosanoids, including its ability to reduce fever and inflammation, increase the rate of gastric secretion and delay the onset of labour.

Prostaglandins (and PG agonists, which have enhanced stability) also have clinical applications. Their most controversial use has been the induction of premature labour, facilitating abortion without the need for a surgical procedure. Some physiological effects of the eicosanoids are given in Table 4.2.

**Table 4.2**   Some physiological effects of eicosanoids.

| |
|---|
| Regulation of platelet aggregation |
| Vasodilation |
| Contraction of uterine smooth muscle (important during labour) |
| Broncho-constriction in the lung |
| Inhibition of lipolysis in adipose tissue |
| Regulation of acid secretion in the stomach |
| Regulation of water reabsorption in the kidney |
| Modulation of synaptic transmission |
| Regulation of lymphocyte activity (immune system) |
| Stimulation of inflammation (released from mast cells) |

### 4.4.4   Growth factors

At this point it is worth clarifying the nature and role of a broad class of chemical mediators described under the general heading of growth factors. In the strict meaning of the term these are polypeptides that stimulate DNA synthesis and cell proliferation following interactions with specific receptors that are present on target tissues. By convention, the term 'growth factors' also includes polypeptides that cause differentiation. You will recall a number of these chemical substances from the previous chapter (Section 3.7.4); examples include epidermal growth factor (EGF), insulin-like growth factor I (IGF I), nerve growth factor (NGF) and platelet-derived growth factor (PDGF). Certain growth factors are also classified among the cytokines (Section 4.4.1) on the basis of their role in activating cell division and differentiation in the immune response to infection: an example is granulocyte–macrophage colony-stimulating factor (GM-CSF).

▷   What is the common feature shared by the receptors of the above-mentioned growth factors and why is it important?

▶   They all exhibit intrinsic tyrosine kinase activity following binding by the appropriate ligand and this contributes to the intracellular signal transduction pathway (Chapter 3).

Growth factors may be either local or endocrine hormones. For example, IGF I (sometimes called a 'somatomedin') is often regarded as an endocrine hormone, since it has been shown to be produced in large quantities by the liver of young animals in response to growth hormone (GH) – see Table 4.1 for additional information on GH. (The liver could, in this sense, be thought of as a 'gland'.) Once released, IGF I is transported in the bloodstream bound to specific proteins which serve to deliver the growth factor to its receptors on target cells, such as the long bones of the skeleton, where it stimulates growth. However, the liver is not the only source of IGF I in the body. Many other cell types also synthesize and release the growth factor, although in smaller amounts. This is indeed a major distinction between growth factors and endocrine hormones: growth factor synthesis and release is not restricted to a specialized gland. The IGF I produced in this way tends only to exert effects in the vicinity of its site of release and so acts rather like a local hormone, activating proliferation in neighbouring cells and also self-stimulating (via an autocrine loop). Such local effects are a common feature for many growth factors, such as NGF, PDGF and EGF. However, these factors can also exert longer range effects, being distributed through the bloodstream.

## Summary of Section 4.4

Locally acting hormones subserve both autocrine and paracrine functions. They include the cytokines and eicosanoids and gases such as nitric oxide. Like endocrine hormones, these ligands exert their effects via specific receptors, but the compounds are much less stable and do not rely on the circulatory system as a means of transportation. A range of stimuli can evoke the immediate synthesis of these local hormones; they can be produced in a diverse range of tissues. The enzymes involved in local mediator synthesis and their receptors can be manipulated using drugs and this can have clinical benefits in the treatment of a number of diseases.

Cytokines are short-range peptides synthesized and secreted by cells involved in coordinating and targeting an immune response. They include several classes of molecule with a range of functions, for example activation of the target cell to secrete cytokines of its own or to proliferate and differentiate, inflammatory reactions around sites of infection, and antiviral effects. Eicosanoids are a family of hormones derived from 20-C polyunsaturated fatty acids with a diverse range of physiological properties ranging from involvement in the pain response to the induction of labour. Nitric oxide is the first to be identified of a group of rapidly diffusing gaseous signalling molecules, active in the nervous system in ways which will be more fully discussed in the next section. Growth factors are a family of polypeptides produced by a diverse range of cell types (rather than a gland). Their growth-promoting effects are often localized, but may extend over a longer range by transport in the blood; like the other short-range signals described, they are mediated by specific receptor systems.

## 4.5   Communication across fixed 'gaps' between cells

In the adult organism cells in any organ or tissue are relatively fixed in position with respect to their neighbours. When a cell dies, it is normally replaced by another of the same type (except in the adult nervous system, where neurons are not replaced). Tissue, you will recall, implies a group of cells of identical type, whereas in an organ cells of different types are located (for example, the neurons and glia of the brain, the $\alpha$ and $\beta$ cells of the pancreas). But in both tissues and organs the arrangement of cells needs to be preserved, whilst it is also often necessary for cells to communicate with their neighbours directly so as to ensure coordinated action. Cellular arrangement is maintained, as you will remember from Book 1 and earlier in this book, by means of glycoprotein adhesion and recognition molecules on the cell surface, whilst the simplest forms of intercellular communication between neighbours is ensured by the presence of intercommunicating gaps in the membranes – so called gap junctions.

### 4.5.1    Adhesion molecules

A variety of glycoproteins present in the cell membrane are involved in physical cell-to-cell associations and also in interactions with the extracellular matrix. The binding that occurs between these cell surface molecules on different cells is very specific, and highly analogous to the many ligand–receptor interactions that you have encountered in this and the previous chapter.

The different sorts of cell surface glycoproteins typically perform two key functions:

1    They promote adhesion and recognition, holding cells together or attaching them to the extracellular matrix.

2    They may also facilitate some form of intracellular signalling following the activation of transduction systems at the cell membrane.

The adhesive function of the glycoprotein cell adhesion molecules results from the fact that the sugar residues protrude from the extracellular domains of the molecules, enabling either 'like-with-like' or 'like-with-unlike' binding between the molecules on the surface of adjacent cells or of the extracellular matrix (Figure 4.18).

This latter process is particularly important in development, where the newborn cells must migrate from the zone in which they are produced to their position in the adult organism; *substrate adhesion molecules* within the extracellular matrix form a sort of slimy snail-trail along which the migrating cells can route themselves. The transduction function of the adhesion molecules arises from the possibility of ligand binding to their extracellular domains serving receptor-type functions and modulating intracellular processes. A wide variety of these intercellular communication molecules have also been identified on the surface of many cells of the immune system. The glycoprotein interactions are particularly important in localizing an immune response, recruiting cells to a specific area of infection.

The magnitude of such intercellular interactions may be regulated by controlling the expression of these types of glycoproteins on the surface membrane of various cells (that is, at a genetic level). In addition, the affinity of these molecules for their corresponding ligands can also be altered by the addition of phosphate groups to the intracellular 'cytosolic tail' region of such proteins.

### 4.5.2    Gap junctions

**Gap junctions** are channels that permit the movement of small water-soluble molecules between adjacent cells. They are widely distributed in all animal tissues and are an important way of connecting cells electrically (through the movement of ions) and metabolically (through the movement of substrates).

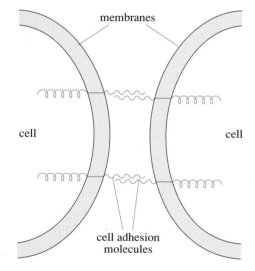

**Figure 4.18**  Schematic representation of cell recognition mediated by cell adhesion molecules. The binding of 'like-with-like' molecules is shown.

At a gap junction the membranes of adjacent cells consist of protein channels called *connexins* that traverse entire lipid bilayers. Each channel is formed from six connexin subunits: these adopt a circular organization which represent the circumference of the pore (Figure 4.19). In order for communication between adjacent cells to take place, two channels (each synthesized by different, neighbouring cells) align to create a functional pore, so that the aqueous environments of two cells become continuous.

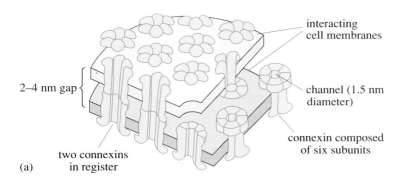

interacting
cell membranes

2–4 nm gap

channel (1.5 nm
diameter)

connexin composed
of six subunits

two connexins
in register

(a)

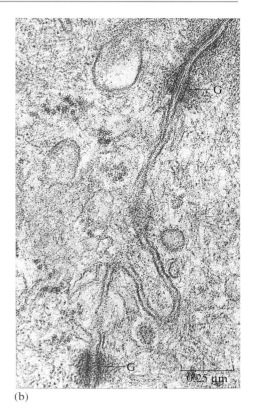

(b)

**Figure 4.19** (a) Schematic diagram of a gap junction. (b) Electron micrograph of cell membranes that are joined by gap junctions (G).

These aqueous contacts can be visualized by microinjecting small fluorescent molecules directly into the cells. After a short time fluorescence can be detected in neighbouring cells, although none is present in the extracellular space. The implication is that the fluorescent compound must have been passed from cell to cell without leakage. Such studies also helped to establish the size threshold for molecules that are able to move between cells via gap junctions: fluorescent compounds with $M_r$ values above 1 000 do not diffuse beyond their cell into which they have been injected, presumably because of the limited dimensions of the pore. However, there does not appear to be any constraint on the type of substance that moves through such gap junctions other than that of relative molecular mass: ions, sugars, amino acids, nucleotides and second messengers such as cAMP, for instance, all readily diffuse between adjacent cells.

The importance of junctions of this sort is that they permit close communication between cells within a tissue, thus for instance ensuring coordinated activity. In heart muscle cells, the flow of ions through adjacent cellular channels permits the transmission of the action potential, allowing the synchronous contraction of cells to generate a heart beat (discussed further in Chapter 7). Similarly, peristaltic movements of the gut (important during digestion) are coordinated by cell-to-cell communication in contracting smooth muscle tissue.

Metabolic communication is also important, particularly for the nourishment of cells that lie some distance from blood vessels, such as lens and bone tissue. Gap junctions have also been implicated in embryogenesis, since intercellular pores could provide a means of conveying substances (termed *morphogens*) which control the development of cells.

There is evidence to suggest that gap junctions are not constitutively open, but may be subject to regulation. The best documented example of this is the closing of intercellular channels in an intact healthy cell when its neighbours are seen to die. The isolation of cells by a similar process may also be important during cell differentiation.

### 4.5.3  Plasmodesmata

**Plasmodesmata** are structures analogous to gap junctions that are present exclusively in plants. These channels are a particularly important route for intercellular communication since plants do not possess a nervous system for the rapid transfer of information. However, their structural organization is different to that of a gap junction. Recall that all plant cells are surrounded by an outer wall in addition to the external cell membrane. Cell walls are, potentially, a major barrier against intercellular interactions of any description, but many of these problems are overcome by the presence of gaps in the walls enabling a continuous aqueous pathway to be formed through the cell membranes between neighbouring cells (Figure 4.20). There is evidence to suggest that the movement of solutes through these structures is selectively regulated. We return to the subject of plasmodesmata in the next chapter.

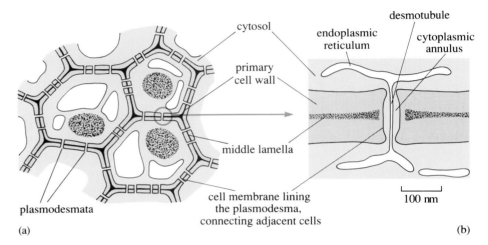

**Figure 4.20**  The structural organization of plasmodesmata. (a) The plasmodesmata are cytoplasmic channels that pierce the cell wall and connect cells in the plant together. (b) The plasmodesma is lined with plasma membrane common to two connected cells, and usually contains a fine tubular structure, the desmotubule, derived from endoplasmic reticulum.

### 4.5.4  Synaptic transmission

Finally in this chapter we turn to the most specific form of chemical transmission of all – that occurring at the synaptic junctions between individual nerve cells and between nerve cells and their effector organs such as muscles. As you will discover, you have already met some of the chemical participants in this process, the neurotransmitters, and the general principles involved in neural signalling will prove familiar from what has gone before in this and the previous chapter.

As you know, the propagation of the nervous impulse, the signal which traverses the dendrites that collect incoming signals, and the axons which then transmit those signals further, is carried electrically, by the flux of ions across the axonal membrane which constitutes the *action potential*, and the detailed mechanism by which this potential is propagated forms one of the subjects of the following chapter; here we are concerned with what happens when the action potential arrives at the synapse and the signal must cross the 0.5 μm gap between the pre- and post-synaptic neurons. For many years after physiological and anatomical studies had revealed the presence of synaptic junctions between nerve cells it was believed that transmission across the synapse was itself electrical, and that the junction was a tight one, rather similar to the gap junctions described in Section 4.5.2. The first neurotransmitters were in fact discovered in the 1920s, but were assumed to operate only in the peripheral nervous system and at the neuromuscular junction, whereas within the brain signalling was still supposed to be electrical.

It was not until the 1950s that the physiological and pharmacological evidence for chemical signalling within the brain became overwhelming.

The first neurotransmitters to be identified were substances you have already met, acetylcholine and adrenalin (and in the brain the closely related noradrenalin). Again, simplicity was the order of the day and it was originally thought that these were the only neurotransmitters and indeed, that each neuron contained only one type of transmitter. Alas for such simplicity; the list of neurotransmitters now runs to several dozen, neurons may contain more than one type of transmitter and, further, to each transmitter there may correspond many different types of receptor. Furthermore, to the list of neurotransmitters must be added other substances which signal within the nervous system, including **neuromodulators** and **retrograde messengers.**

Many of these substances are closely related chemically to some of the hormones discussed in Section 4.2. This cannot be coincidental, and one plausible hypothesis is that hormonal signalling developed early in evolutionary history, probably along with the first multicellular organisms. Neurons evolved as a way of providing specific rather than general signalling to distant cells within the body, and utilized the existing hormones and receptors as their mode of communication; in a sense chemical transmission at the synapse is no more than a specialized type of paracrine process.

A typical chemical synapse within the brain is shown in the electron micrograph of Figure 4.21.

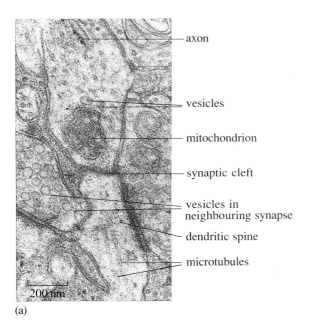

(a)

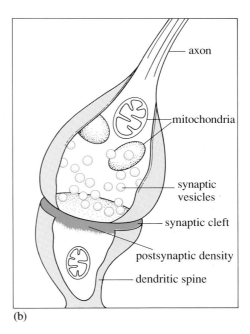

(b)

*Figure 4.21* (a) Electron micrograph of a synapse within the brain. (b) Schematic diagram of the synapse shown in (a).

Figure 4.21 shows the swelling at the termination of the axon, containing mitochondria, which provide the energy source, and densely packed **synaptic vesicles**, small spherical granules which contain the neurotransmitter. Any given neuron may make between 10 000 and 100 000 synaptic contacts with the dendrites of other neurons, either neighbouring or distant. During synaptic transmission, as a result of the depolarizing signal that arrives down the axon by way of the action potential (discussed in the next chapter), the vesicles move to the junction itself, that is, the thickened area visible in the electron micrograph, where they fuse with the synaptic membrane. The guidance for this process is provided by the microtubules that run down the axon and traverse the synaptic swelling to finally be anchored to the junction by way of the proteins *fodrin* (described in Book 1) and *spectrin*. At the junction the vesicles empty their contents into the **synaptic cleft** which lies between the pre- and post-synaptic membranes. The released neurotransmitter diffuses across the cleft to the post-synaptic side where it is bound to the receptors embedded in the membrane. The extracellular space between the pre- and post-synaptic membranes is fluid filled, but also contains the extracellular domains of the glycoprotein neural cell adhesion molecules which stretch across the cleft from either side, adhering to one another so as to hold the synapse in position.

The binding of the neurotransmitter to its receptor opens channels in the post-synaptic membrane allowing the entry of ions by mechanisms you are familiar with from this and the previous chapter. Neurotransmitters are of two general types: excitatory and inhibitory. The binding interaction of excitatory transmitters permits the entry of cations, especially sodium, and for some classes of neurotransmitter, calcium, resulting in a depolarization of the post-synaptic cell and, if the depolarization is sufficient, the generation of a potential which spreads along the post-synaptic dendrite into the cell body of the neuron. Here, in interaction with other such potentials arriving from the many other synapses and dendrites, it may summate to generate an action potential down the axon. Inhibitory transmitters by contrast open chloride channels, which result in **hyperpolarization** and hence tend to inhibit the propagation of an action potential.

The prototype excitatory neurotransmitter in the central nervous system is the amino acid *glutamate*, and the inhibitory neurotransmitter *GABA* (*gamma-aminobutyric acid*), though there are many others in each category. Each neurotransmitter is capable of interacting with several different post-synaptic receptors. Thus there are four general types of glutamate receptor, each of which has a number of subclasses, some associated with fast and transient, some with longer-lasting responses in the post-synaptic cell. Several of these have now been cloned and their structure determined.

Particular neural pathways within the central nervous system, in which signals may cross several synapses, often all utilize the same neurotransmitter, and amongst those of major significance are included the amino acid derivatives *serotonin* (5-hydroxytryptamine) which is important in pathways concerned with attention and emotion, and *dopamine*. The 1970s and 80s saw the discovery of a whole new class of peptide neurotransmitters, in particular those implicated in pathways concerned with pleasure and pain, the *enkephalins* and *endorphins*. In addition peptide fragments derived from the hormones vasopressin and adrenocorticotropin were also shown to have neurotransmitter or neuromodulator activity.

Because the essence of neuronal as opposed to hormonal signalling is rapid onset and offset, once the neurotransmitter/receptor interaction has occurred and the relevant depolarization or hyperpolarization has followed, it is important that the channel be closed, switching off the signal. The transmitter must be removed from the receptor so that it becomes available to receive signals once more. A number of mechanisms are available for the removal of transmitter. These include destruction by specific enzymes (for example, *acetylcholinesterase* for acetyl-choline) or re-uptake by way of another set of specific receptors into the pre-synaptic side of the synaptic junction, or into the many non-neuronal cells (the glial cells) which surround the synaptic junction.

The delicate stages involved in this process means that signals traversing the synaptic cleft are particularly vulnerable to disruption by drugs and poisons.

▷ How many mechanisms can you think of which would result in such disruption?

► Here are some:

(a) **Agonist** molecules that mimic the transmitter but cannot be subsequently destroyed or removed.

(b) **Antagonists** that compete for the receptor.

(c) Substances that prevent the destruction of the neurotransmitter after signalling has been completed, either by inhibiting the enzymes which destroy it or by blocking re-uptake mechanisms.

The results of such disruption vary depending on the nature of the neural pathway whose messages are being thus intercepted. For example, the transmitter at the neuromuscular junction is acetylcholine acting in its nicotinic mode and the result, depending on whether its action or its destruction is blocked, is either muscular paralysis or rigid contractions. Many natural poisons such as snake venoms and toxins act this way; so do chemicals such as organophosphorus compounds which inhibit acetylcholinesterase – these form the basis both of some insecticides and the toxic nerve gases of chemical warfare.

A number of clinical conditions are also associated with disruption of particular neurotransmitter systems, and here pharmacologically produced drugs can alleviate some of the symptoms of the condition. The characteristic tremors of Parkinson's disease, for instance, are due to the destruction of the dopamine-producing cells of a particular brain region, the basal ganglia. The lack of this neurotransmitter prevents the neural feedback loop via the basal ganglia and a region of the brain called the cerebellum, which normally regulates and smooths motor activity. Treatment with a chemical related to dopamine, *L-dopa*, can provide some relief. The mental distress of depressive disorder is associated with a failure of the serotonergic pathways, and drugs such as *imipramine* and *paroxetine*, which bind to and block the serotonin re-uptake mechanisms are widely prescribed as antidepressants. *Benzodiazepine*, the active component of drugs such as the tranquilizer or anxiolytic valium, is believed to interact with GABA receptor sites in the brain, although it is not clear that this is its only site of action. And the pain-relieving effect of *morphine* had been known for centuries before it was recognized to be an analogue of the peptide endorphins.

As well as neurotransmitters we have already referred to neuromodulators and retrograde messengers. Neuromodulators are really the nervous system's version of paracrine hormones. Rather than act directly and specifically through synaptic signalling, they are released by neurons to diffuse through the extracellular space between cells, where they interact with receptors so as to either potentiate or diminish their receptivity to the specific neurotransmitters.

Retrograde messengers have a more specific function. They are produced in the post-synaptic cell in response to receptor activation and released into the synaptic cleft, where they diffuse back to the pre-synaptic cell so as to signal to it that the post-synaptic cell has responded. This is particularly important for the phenomenon of *synaptic plasticity*, the modulation of synaptic connectivity by making or breaking synaptic contacts between cells. This process is important during development, when the effect of environmental experience helps to shape the pattern of synapses, and is also believed to occur during learning and memory formation. Memories which result in changed patterns of behaviour are believed to be stored in the brain as a result of changes in synaptic connectivity. The mechanism by which this is achieved requires coordinated remodelling of pre- and post-synaptic neurons and their synaptic connections. Two molecules you have already met in the previous section, nitric oxide and arachidonic acid, are believed to act as retrograde messengers in this type of signalling process during memory formation. They are released when calcium enters the post-synaptic cell through channels opened by the interaction of the neurotransmitter glutamate with one of its receptors, called the NMDA (*N*-methyl-D-aspartate) receptor.

▷  Can you think of an experiment which might test this hypothesis?

▶  If you were to train an animal on a particular task, and then to block the synthesis of nitric oxide or arachidonic acid, the animal should be unable to remember the training it had learned.

In fact this type of experiment has been done. In 1992 Christian Holscher and Steven Rose at the Open University showed that blocking NOS with nitroarginine (analogous to the monomethylarginine referred to in Section 4.4.2 above) prevents memory for a simple learning task in young chicks; there was a similar result when a metabolic blocker of arachidonic acid synthesis (the substance nordihydroguaiaretic acid) was used. At the time of writing this Course the role of these substances and other eicosanoids in the control of nervous system function is only beginning to be explored, and there are certainly many more exciting results to come during the lifetime of S327. The interactions of neurotransmitters, receptors, neuromodulators and retrograde messengers in chemical signalling at the synapse are summarized in Figure 4.22 .

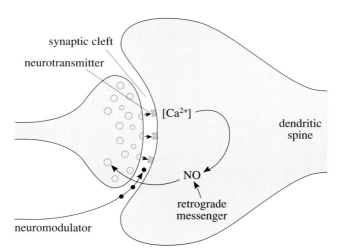

**Figure 4.22**  Interactions of neurotransmitters, neuromodulators and retrograde messengers.

## Summary of Section 4.5

Order and recognition between adjacent cells in tissues and organs and with the extracellular matrix is provided by the cell adhesion molecules. These also serve to transduce signals between cells and during the immune response. The presence of gap junctions in many animal tissues permits the direct movement of small water-soluble molecules between adjacent cells, providing a pathway for electrical and metabolic communication. Aqueous contacts (called plasmodesmata) found between adjacent plant cells serve a similar function in local communication, but exhibit greater selectivity of solute movement.

Chemical transmission at the synapse shares many features in common with hormonal signalling, especially of paracrine hormones. Transmitters are released from vesicles in the pre-synaptic axon termination and cross the synaptic cleft to bind to post-synaptic receptors, where the consequence is the opening of ion channels. Release of excitatory transmitters such as glutamate results in the entry of cations and post-synaptic depolarization; inhibitory transmitters such as GABA result in entry of chloride and hyperpolarization. There are at least several dozen different neurotransmitters, as well as neuromodulators. Retrograde messengers are released from the post-synaptic cell and diffuse back to the pre-synaptic side to generate responses in the pre-synaptic neuron; these include nitric oxide and arachidonic acid. A variety of substances – agonists and antagonists, toxins and enzyme inhibitors – can interfere with these signalling processes. In some cases, such as that of transmission by acetylcholine at the neuromuscular junction, the result may be death by muscular paralysis or contraction. In others, drugs that interact with central nervous system neurotransmission may affect learning and memory, attention and arousal, pleasure and pain; they may produce altered states of consciousness and alleviate the symptoms of mental distress such as depression and anxiety, and neurological diseases such as Parkinson's.

## Objectives for Chapter 4

After completing this chapter you should be able to:

4.1    Define and use, or recognize definitions and applications of, each of the terms printed in **bold** in the text.

4.2    Compare signalling properties in the endocrine, immune and nervous systems, including their structural organization and the timing of their responses.

4.3    Outline the sequence of events which follows the activation of intracellular and membrane-bound surface hormone receptors, and explain the observed differences in the time taken to generate a biological response.

4.4    Describe the regulation of hormone concentration, including control by feedback inhibition.

4.5    Explain the ways in which biological responses to hormones can be modulated and how endocrine disturbances may arise.

4.6    Illustrate long-range and short-range chemical signalling in the immune system, using the example of the antibody response to an antigen.

4.7    Provide examples of some physiologically important local mediators and describe how their synthesis can be regulated.

4.8    Explain possible therapeutic benefits which arise from the activation or inhibition of local mediator systems through the use of specific drugs.

4.9    Identify the gap junctions and plasmodesmata structures in cells, and explain their physiological importance in short-range intercellular communication.

4.10    Describe the sequence of events occurring when an action potential arrives at an excitatory synapse.

## Questions for Chapter 4

### Question 4.1  (Objectives 4.1 and 4.10)

Although the action potential is the key feature of electrical signalling in the nervous system, most neuronal pathways also involve chemical signalling. Explain.

### Question 4.2  (Objectives 4.2 and 4.10)

Why is it not possible to signal across a synapse in both directions?

### Question 4.3  (Objective 4.4)

The synthesis of thyroid hormones requires iodine, which is incorporated into thyroid hormone molecules. In some remote inland areas (such as the Andes region in South America) there is a lack of iodine in the diet and this may result in an impairment of thyroid hormone synthesis. What would you expect to happen to the plasma TRF and TSH concentrations of a thyroid-deficient individual? (Figure 4.6 will help.)

### Question 4.4  (Objective 4.5)

Alcohol is well-known to increase water excretion and urine flow (diuresis). Water homeostasis is primarily controlled by the secretion of vasopressin (antidiuretic hormone) from the pituitary gland. The hormone acts on the kidney, promoting water reabsorption (Table 4.1). What are the possible explanations for the observed effect of alcohol on water balance?

### Question 4.5  (Objective 4.6)

Briefly describe a cell–cell interaction in the immune system involving receptor–ligand binding between molecules on the surfaces of communicating cells, together with short-range chemical signalling.

### Question 4.6  (Objectives 4.1 and 4.7)

What properties make nitric oxide an effective paracrine or autocrine mediator?

### Question 4.7  (Objective 4.9)

Which of the following statements do not apply to gap junctions?

(a)  They allow metabolites to reach cells which are found some distance from blood vessels.

(b)  They are formed when two protein 'half-channels' align to form a continuous aqueous pathway between adjacent cells.

(c)  Charged molecules move through gap junctions more readily than uncharged substances.

(d)  The protein pores that connect cells can be regulated (i.e. opened and closed).

### Question 4.8  (Objectives 4.1 and 4.7)

What is the source of N for nitric oxide synthesis? (a) Citrulline; (b) urea; (c) arginine; (d) ammonia.

*Question 4.9*  *(Objectives 4.2 and 4.7)*

Compare and contrast the properties and mode of action of endocrine hormones and locally acting mediators.

*Question 4.10*  *(Objectives 4.7 and 4.8)*

Two classes of drug, steroid (e.g. cortisol) and non-steroid (e.g. aspirin, indomethacin), are currently used for the treatment of inflammatory diseases such as rheumatoid arthritis. What are the similarities and differences between the way that these drugs act?

## CHAPTER 5

**_Electrical signalling between cells_**

## 5.1   Introduction

In the previous chapter we considered signalling and communication within an individual cell and then went on to look at the signalling systems within a multicellular organism (typically a mammal) – systems which help to preserve the homeostatic and homeodynamic integrity of that organism. As we said in the introduction to Chapter 4, such signalling systems are of three types: endocrine, immune and nervous. Unlike the chemical modes of transmission in the endocrine and immune systems, however, the major mechanism of signal propagation within the nervous system is electrical. To understand how this mechanism works, it is necessary to go back to some of the principles developed in Chapter 2 (again, bear in mind the health warning concerning the equations, and remember that you do not need to recall mathematical derivations or to reproduce the equations except as indicated in the Objectives for the current chapter). We can give only a brief account of the processes involved; for a more detailed description, you could refer to a good neurophysiology textbook or to the Open University course SD206, _Biology, Brain and Behaviour_.

You will recall from Chapters 2 and 3 how ion channels serve to convey electrical and chemical messages into the cell across the external cell membrane. Once across the membrane, the messenger ions may diffuse around the cell. Remember that the average time that an ion takes to diffuse a distance $r$ is proportional to $r^2$, so diffusion is quick and efficient over the short distances _within_ most cells but it would be very slow and inefficient over the much larger distances required to communicate _between_ cells.

What does this mean in practical terms? It was shown in Chapter 2, Equation 2.1, that the average distance, $r$, that an ion with a diffusion constant $D$ will diffuse in a time $t$ is given by

$$r = \sqrt{(6Dt)}$$

Thus a $Na^+$ ion with a diffusion constant of $1.33 \times 10^{-9}\,m^2\,s^{-1}$ would take, on average, 12.5 ms to diffuse across a cell of diameter 10 μm. So, ions admitted to a cell by an ion channel that opens through the external cell membrane could make their presence felt throughout that cell in a time of the order of 10 ms. However, if the distance across which the signal is to be transmitted is larger, the time required for transmission increases very considerably. In animals, electrical signals are conveyed between the brain and the limb extremities; in larger creatures this can be over distances of the order of 1–1.5 m. A $Na^+$ ion would take, on average, 1 253 days to diffuse 1 m, and this would result in a very slow animal indeed!

So what means are employed by living organisms to transmit electrical messages over long distances? Neurons, which are in effect specialized nerve cells, provide the major mechanism, conveying electrical messages over large distances in a few milliseconds. The greater speed is due to the harnessing of energy that is stored as voltage and ionic gradients. But before we discuss neurons, we shall consider a simpler means of communication between neighbouring cells that is important in animals without nervous systems and may even operate in plants – namely chain propagation or short-range electrical coupling.

## 5.2   Chain propagation – a short-range electrical coupling mechanism

Consider a chain of similar cells with the cytoplasm of each cell connected to that of the neighbouring cell on either side of it by gap junctions which, as you read in the previous chapter, span the two adjacent cell membranes where they touch. To study the electrical behaviour of such a chain as simply as possible, we shall use the electrical model of the cell that was introduced in Section 2.3 and is illustrated in Figure 2.6. In this model, the cytoplasm of each cell is represented by a body of conducting fluid at the same voltage, $V_{in}$, throughout. It is separated by a non-conducting membrane from the fluid bathing the cell; that fluid is at the voltage, $V_{out}$. Then the only voltage difference to which the cell can respond is given by the membrane voltage, $V_m$, where

$$V_m = V_{in} - V_{out}$$

As the conducting fluid bathing the outside of the cells has a greater volume than any of the cells, it will also have a greater electrical capacitance than any of the cells. This means that when a small amount of charge is transferred between a cell and the bathing fluid, across the cell membrane, the change in $V_{out}$ will be very much less than the change in $V_{in}$, so the changes in $V_{in}$ become the same as changes in $V_m$ – in other words, $V_m$ and $V_{in}$ can be treated as interchangeable.

We shall suppose (see Section 2.5.1 and Figure 5.1 below) that each cell membrane is spanned by voltage-gated ion channels that are selective for a univalent cation $X^+$. This cation has a higher free concentration outside the cells than inside so that the Nernst potential, $V_{N(X^+)}$, of the cation is positive. The gated ion channels open when the membrane voltage, $V_m$, is more positive that some critical voltage, $V_c$. When at rest, the voltage of the cytoplasm of all the cells that are electrically connected together by the gap junctions, is $V_r$.

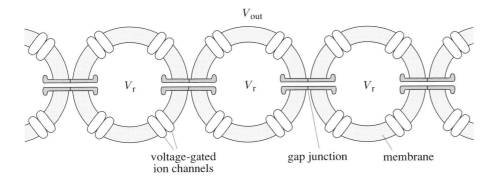

voltage-gated
ion channels          gap junction          membrane

**Figure 5.1**   A simplified picture of a chain of cells connected together by gap junctions (a single representative junction is shown between each pair of cells) with voltage-gated ion channels spanning their membranes. The cytoplasm of each cell rests at a voltage, $V_r$, while the conducting fluid that surrounds the cells is everywhere at a voltage, $V_{out}$.

If positive charge is injected into any cell in the chain, its internal voltage will rise until its membrane voltage exceeds $V_c$. Then the ion channels of that cell will open, $X^+$ ions will flood into the cell, and its membrane voltage will rapidly rise further until $Vm = V_{N(X^+)}$ and the $X^+$ ion is in equilibrium across the cell membrane. Because it now has a higher internal voltage higher than the neighbouring resting cells ($V_{N(X^+)} > V_r$), positive charge will flow from this cell into its two neighbours in the chain through the gap junctions. The internal voltages of each of the two neighbouring cells will then rise until their membrane voltages reach the critical value $V_c$, whereupon their voltage-gated ion channels will open and their internal voltages will rise rapidly to $V_{N(X^+)}$. Current will then flow to the next cells in the chain, raising their internal voltages, and so the initial excitation of a single cell will be propagated along the chain of connected cells. Such a propagating electrical potential wave is known as an action potential and its transmission along the chain of cells is an example of **signal propagation.** Note that if a single cell in a chain is excited, the signal will propagate in both directions along the chain of cells away from the excited cell.

We shall now discuss in more detail the time-course of the voltage changes by examining the passage of an action potential along just three cells in a connected chain as shown in Figure 5.2.

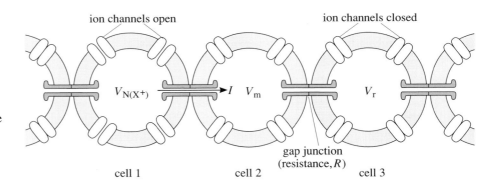

**Figure 5.2** An electrical model of three cells in a connected chain showing the current, $I$, flowing from a previously excited cell into cell 2 through gap junctions with a resistance, $R$.

We shall concentrate on the voltage changes within cell 2 (remember that $V_{in}$ and $V_m$ are interchangeable). We shall assume an initial state at time $t = 0$ when cell 1 has already been excited and its voltage has reached $V_{N(X^+)}$, while cells 2 and 3 still have their resting voltages, $V_r$. An electric current, $I$, now flows from cell 1 to cell 2 through the connecting gap junctions which are assumed to present a resistance, $R$, to the current flow. Using Ohm's law, we now write

$$V_{N(X^+)} - V_m = RI \qquad \text{or} \qquad I = (V_{N(X^+)} - V_m)/R \qquad (5.1)$$

The current $I$ will vary with $V_m$, the membrane voltage of cell 2, becoming smaller as $V_m$ rises. However, as a first approximation, we shall assume that when $V_m = V_r$, the current $I$ is constant at the initial value it had at time $t = 0$, so that

$$I = (V_{N(X^+)} - V_r)/R \qquad (5.2)$$

Electric current is just the rate of flow of charge, so a current $I$ flowing for a time $t$ delivers a total charge $Q$, where

$$Q = It \qquad \text{or} \qquad t = Q/I \qquad (5.3)$$

To estimate the rate of rise of the membrane voltage of cell 2 as positive charge flows into it, we shall consider the cell to be an electrical capacitor with a capacitance, $C$, as we did in Chapter 2, Section 2.3. To charge a capacitor of capacitance $C$ to a voltage $V$ requires a charge $Q$ where

$$Q = CV$$

Substituting this value for $Q$ into Equation 5.3 and using the value of $I$ from Equation 5.2, we obtain an expression for the time, $t_c$, required for the current $I$ from cell 1 to raise the membrane voltage of cell 2 to the critical value $V_c$ when its ion channels open to admit $X^+$ ions:

$$t_c = RC[(V_c - V_r)/(V_{N(X^+)} - V_r)] \qquad (5.4)$$

A more accurate model, which allows for the reduction of the current, $I$, as the membrane voltage of cell 2 rises from $V_r$ towards $V_c$, leads to the equation

$$t_c = RC \log_e[(V_{N(X^+)} - V_r)/(V_{N(X^+)} - V_c)] \tag{5.5}$$

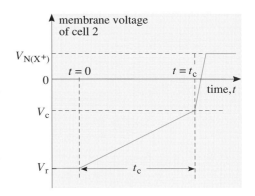

After the membrane voltage of cell 2 has reached the critical value $V_c$, and its ion channels have opened allowing the rapid inflow of $X^+$ ions, its membrane voltage very quickly reaches the Nernst potential, $V_{N(X^+)}$, for the $X^+$ cations. The time-course of these changes in the membrane voltage of cell 2 is sketched in Figure 5.3. It can be seen that by far the greatest contribution to the time required for cell 2 to increase its membrane voltage from the resting value, $V_r$, to $V_{N(X^+)}$, is the time $t_c$. In both Equations 5.4 and 5.5, $t_c$ is proportional to the product $RC$ which is called the time constant, $t_{RC}$, for the charging of the cell.

**Figure 5.3** Time course of changes in the membrane voltage of cell 2 in the connected chain of cells shown in Figure 5.2.

It is interesting to consider the magnitude of $t_c$ and of $t_{RC}$. If we assume that the critical voltage, $V_c$, is midway between the resting voltage, $V_r$, and the Nernst potential, $V_{N(X^+)}$, so $(V_c - V_r) = 0.5(V_{N(X^+)} - V_r)$, then the two equations (5.4 and 5.5 ) give values for $t_c$ of $0.5\,RC$ and $0.69\,RC$ respectively. What does this mean in numerical terms? A typical cell membrane has a capacitance per unit area of about $5 \times 10^{-3}\,\mathrm{F\,m^{-2}}$, so for a 10 μm diameter spherical cell the capacitance, $C$, is $1.57 \times 10^{-12}\,\mathrm{F}$ (F denotes farad units). Suppose that a current of $1 \times 10^{-12}\,\mathrm{A}$ passes through a single gap junction in the membrane when the electrochemical driving force across that junction is equivalent to a voltage of 50 mV, then the resistance of the junction ($R = V/I$) is $5 \times 10^{10}$ ohm. If there are one hundred such parallel junctions connecting adjacent cells, then $R$ is reduced by a factor of one hundred and becomes $5 \times 10^8$ ohm. By definition $t_{RC} = RC$, so with the values of $R$ (for a hundred parallel junctions) and $C$ just estimated, $t_{RC} = 0.785$ ms. The time, $t_c$, approximates to $0.69\,RC$, i.e. $0.69\,t_{RC}$. Therefore, $t_c$ is about 0.54 ms, which corresponds to a signal propagation speed of 10 μm in 0.54 ms, or 18 mm s$^{-1}$, along the connected chain of cells.

Three major advantages of signal propagation by electrical coupling rather than by free ionic diffusion are readily apparent. The first is its effectiveness for signal transmission over relatively long distances: the speed is constant and hence the distance travelled is simply proportional to the time that has elapsed, whereas diffusion is fast over very small distances but is very slow over greater distances. The second advantage of chain propagation is that the signal travels along a chain of connected cells and therefore the propagation path is specific. Thirdly, the ions are driven through the connecting gap junctions in the membrane by powerful electric fields rather than being free to perform a non-directed random walk (see Section 2.2.1) as in diffusion, hence the greater efficiency of chain propagation.

However, unlike diffusion which relies only on thermal energy, chain propagation requires an external energy source. This energy is provided by the drop in electrochemical potential experienced by particular ions when they move into a cell across its membrane. As you may recall from Chapter 2, Section 2.4, the change in the electrochemical potential, $g$, of an ion under these circumstances is given by

$$dg_{in} = g_{in} - g_{out} = ze(V_r - V_{N(X^+)}) < 0 \tag{5.6}$$

The difference has a negative value here because $V_r$, the membrane resting voltage, is negative, and $V_{N(X^+)}$, the Nernst potential for the $X^+$ ions, is positive. The energy is stored by means of the ionic concentration differences that have been built up across the cell membranes by primary and secondary ion pumps fuelled by metabolic energy. Using this stored energy each cell acts as a signal repeating station which amplifies the signal and then passes it on to the neighbouring cells.

In animal cells the $Na^+$ concentration is generally maintained at a higher level outside the cell than inside (see, for example, Table 2.2), so the Nernst potential for $Na^+$ is positive. When the $Na^+$ selective channels open, $V_m$ is driven from its negative resting value towards the positive value of $V_{N(Na^+)}$ by the inrush of $Na^+$ ions through the open channels. This makes voltage-gated $Na^+$ channels ideal for the chain propagation of signals.

But to be useful for such propagation, a chain of cells must return to its resting state quickly, ready to transmit the next electrical signal. This recovery is usually achieved by two mechanisms acting together. In the first mechanism, the $Na^+$ channels inactivate (spontaneously close) a short time after they have opened in response to the rise in the membrane voltage . The second mechanism is more complex: $K^+$ channels open in response to the same rise in membrane voltage that triggers the opening of the $Na^+$ channels, but the response time for opening is slower for the $K^+$ channels and so they open when the $Na^+$ channels have passed their peak conduction and have started to inactivate. Because the Nernst potential for $K^+$ in these cells is negative due to the higher concentration of $K^+$ ions within the cell than outside it, the opening of these channels results in a flood of $K^+$ ions leaving the cell. This, in turn, drives the voltage of the cytoplasm negatively towards its resting potential. Meanwhile, the high rate of ion redistribution across the membrane during the short time that a signal is travelling along the chain is offset by the continuous working of the primary and secondary pumps in the cells: these are continuously pumping $Na^+$ ions out of the cells and $K^+$ ions in at a slow rate so as to return the ionic concentrations to their resting values.

Chain propagation of signals has been measured in assemblies of epithelial cells connected by gap junctions in some animals lacking nervous systems, for example, the simple jellyfish, *Siphonomores* and *Hydromedusae*. Propagation speeds are found to be of the order of 2–500 mm s$^{-1}$. More controversial, but of great interest if confirmed, are the experiments described below which point to a similar mechanism in plant cells.

Plant cells do not have gap junctions like animal cells but are connected together by plasmodesmata (see Chapter 4, Figure 4.20) which have solute permeability and electrical conductivities comparable with those of gap junctions. When the cotyledon (seed leaf) of a young tomato plant is wounded mechanically, an action potential propagates across the cotyledon with a speed of a few millimetres per second and eventually spreads along the leaf stem to the first open leaf, where it can be measured by external electrodes. The presence of the action potential at a given location correlates with the local production of particular proteinase inhibitors known as pin 1 and pin 2. (These enzymes make the leaf harder to digest when eaten by a grazing animal and so, after the initial injury, the plant ensures that its other leaves are less palatable by producing these enzymes.) Until recently it was not known whether the signal for pin production was a consequence of the propagation of a primary chemical messenger transported within the phloem system, or whether the action potential was the primary signal. It has been difficult to determine which of these two pathways is responsible because they would both propagate signals at similar speeds. An attempt has been made to distinguish between the two paths by freezing the stem of the cotyledon to 3 °C prior to the mechanical injury. Radioactive tracer studies have shown that such freezing prevents transport in the phloem, and effectively eliminates this pathway. Yet electrical measurements reveal that, despite the freezing process, the action potential detected on the first leaf is unchanged from first observations and pin production is normal (note that the cotyledon was cut off immediately after the passage of the action potential along its stem to prevent other signals which may propagate more slowly). These experiments suggest that the action potential is the primary signal and indicate that long-distance inter-cell electrical signalling along a linked chain of cells may play an important role not only in animals but also in plants.

## Summary of Section 5.2

A chain of cells with the cytoplasm of neighbouring cells electrically connected together by gap junctions can transmit an electrical signal called an action potential along the chain if any cell in the chain has its internal voltage raised sufficiently. Unlike diffusion, this propagation mechanism requires an energy input and will only propagate in the direction of the connected cell chain. The energy required is stored in the ionic concentration differences of certain ions, these differences being created across the cell membranes by membrane-bound primary and secondary ion pumps.

## 5.3  Signal propagation by neurons

Although those invertebrates whose signalling systems probably evolved before those of the ancestors of vertebrates can manage without the specific signalling system provided by nerves, larger animals require rapid and reliable signal propagation over distances of a metre or more depending on body size. As discussed in Chapter 1 and in the introduction to Chapter 4, a message may be sent over large distances by means of neurons, the signal travelling along specific insulated paths that connect only particular cells and do not make contact with the intervening cells. Recall that the axon of a neuron may be 1 μm in diameter and 0.5 m in length, the most extreme ratio between the dimensions of any known cell.

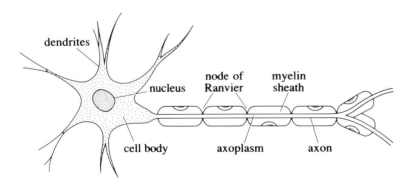

*Figure 5.4*  The structure of a myelinated neuron.

Figure 5.4 will remind you of the salient features of a neuron. In particular, note that the cytoplasm within the axon is known as **axoplasm**. Also note the electrically insulating **myelin sheath** which is wrapped around axons of the peripheral nervous system, and around axons of those neurons of the central nervous system which communicate over relatively long distances. It is the myelin, which is lipid-rich, that gives the 'white matter' of brain its characteristic colour. At regular intervals there are short breaks in the myelin sheath, where the cell membrane of the axon is exposed; these areas are known as the **nodes of**

**Ranvier.** The important biophysical role played by these breaks, as well as the significance of myelin for the speed of signal propagation, will become clear in Section 5.3.2.

As you saw in the previous chapter, neurons communicate with other neurons and muscle cells by the diffusion of neurotransmitters across the synaptic cleft between a pre-synaptic terminal of one neuron and a post-synaptic terminal on the cell body or dendrite of another cell, the release of the neurotransmitters being triggered by the arrival of the action potential. But here we will be concerned largely with the biophysics of signal transmission along the axon rather than communication between neurons.

### 5.3.1 The giant axon of the squid

The properties and mechanism of the action potential have been extensively studied in the giant axon of the squid. This somewhat unusual preparation was focused on in the 1930s, initially by Frank Schmidt in Cambridge, Massachusetts, and by John Z. Young in London. The merits of this axon, from a research point of view, lie in the fact that it is unmyelinated and it has an unusually large diameter of up to 1 mm. It was brilliantly exploited through the 1940s and early 1950s by Andrew Hodgkin and Hugh Huxley in Cambridge, England, and they were eventually awarded a Nobel prize for their elucidation of the biophysics of the action potential.

The $Na^+$, $K^+$ and $Cl^-$ concentrations within the giant squid axon and in the fluid surrounding it are listed below, together with the Nernst potentials of these three ions at 25 °C. The resting membrane voltage, $V_r$, for the axon is about $-70$ mV, which is quite close to the Nernst potentials for $K^+$ and $Cl^-$.

| Ion | Concentration in axon/mmol $l^{-1}$ | Concentration in fluid around axon/mmol $l^{-1}$ | Nernst potential at 25 °C/mV |
|---|---|---|---|
| $Na^+$ | 50 | 440 | +55.9 |
| $K^+$ | 400 | 20 | −77.0 |
| $Cl^-$ | 52 | 560 | −61.1 |

The mode of propagation in the axon is essentially the same as that in the chain of cells we discussed in the preceding section, except that it is much faster because the axon does not have the high resistance of the narrow gap junctions between adjacent cells which slows the flow of ions along the cell chain. Instead of the charge leaking slowly from the excited cell to adjacent cells through gap junctions, a wave of positive charge flows along the cylindrical axoplasm. Near the region of positive charge the membrane voltage of the axon rises and voltage-gated channels selective for $Na^+$ open; as a result $Na^+$ ions rush in and enhance the positive charge in the axoplasm, driving the local internal voltage upward towards the $Na^+$ Nernst potential, $V_{N(Na^+)}$, which is +55.9 mV. As the entering cations diffuse along the axon, they tend to raise the voltage of neighbouring regions of the axoplasm towards $V_c$, causing nearby $Na^+$ channels to open and so the excitation propagates along the axon. As in the cell chain, the internal voltage is subsequently restored to its negative resting value by the inactivation of the $Na^+$ channels and by the delayed opening of the voltage-gated ion channels selective for $K^+$ ions leading to an outflow of $K^+$ ions. The voltage changes in the axoplasm are shown in Figure 5.5: first, there is the abrupt voltage rise due to the influx of $Na^+$ ions, followed by the slower fall due to inactivation of the $Na^+$-selective channels and the outflow of $K^+$ ions through the $K^+$-selective channels. In fact, the open $K^+$ channels tend to drive the internal voltage towards the $K^+$ Nernst potential, $V_{N(K^+)}$, of $-77\,\text{mV}$, slightly overshooting the resting potential, $V_r$, of $-70\,\text{mV}$.

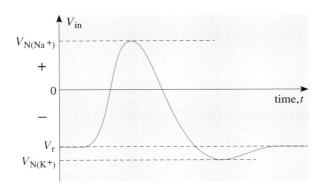

**Figure 5.5** Axoplasmic voltage as a function of time during the passage of an action potential.

### 5.3.2 The axon as an electrical cable

In effect, an axon resembles an electrical signal cable with a large number of repeater stations amplifying the signal as it progresses. As already indicated, the energy needed by the signal amplifiers lies in the $Na^+$ and $K^+$ ionic concentration differences built up and maintained continuously by the cell's ion pumps. The propagation of the signal relies on positive feedback and, in fact, in some ways the underlying mechanism is like that in a burning chemical fuse where the heat generated by the exothermic oxidation at one point is sufficient to initiate the oxidation of the adjacent section of the fuse. The speed of propagation varies with the diameter of the axon and with the degree of myelination as will be discussed later in this section; it ranges from about $1\,\text{m s}^{-1}$ to $100\,\text{m s}^{-1}$, much higher than in a cell chain.

As an extension of the cable analogy, the axon may be compared to a conventional insulated copper electrical cable (see Figure 5.6a overleaf). The electrically conducting axoplasm plays the role of the copper core of the cable, and the axonal cell membrane is analogous to the insulating cable. However, the axon is a very poor cable in comparison with the insulated copper version as it presents a high resistance to the signal and, moreover, much of the signal leaks out through the insulation.

We shall now consider what this analogy can tell us quantitatively. If a piece of electrical cable of length $x$ has a core resistance of $R_c$, then a cable of twice that length has a core resistance of $2R_c$ because the two resistors simply add *in series*. On the other hand, the leakage resistance follows a different formula: if the leakage resistance through the insulation of a piece of cable of length $x$ is $R_l$, then the leakage resistance of the insulation of twice that length of cable is $R_l/2$ because the two leakage resistors are *in parallel*, and so, as there is less leakage resistance, the rate of leakage is greater. Now, thinking in terms of a unit length of cable, we find that, if the resistance per unit length of the core is $R_i$, then the resistance of a section of cable of length $dx$ will be $R_i dx$; while if the leakage resistance of the insulation per unit length of cable is $R_m$, then the leakage resistance of a length of cable $dx$ is $R_m/dx$. Similarly, if the capacitance of the insulation of a unit length of the cable is $C_m$, then the capacitance of the insulation of a section of cable of length $dx$ is $C_m dx$ as the capacitances add in parallel.

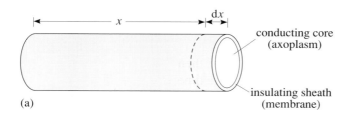

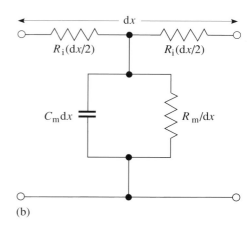

**Figure 5.6** (a) The axon represented as an electric cable. (b) The analogous circuit for a length d$x$ of an axon.

Figure 5.6(b) shows the analogous circuit for a length d$x$ of an axon regarded as a passive electrical cable. It is assumed that, because the conducting fluid bathing the outside of the axon has a volume much larger than that of the axon, it has a negligible resistance; thus currents flowing in the bathing fluid give rise to negligible voltage differences. The resistance of the interior of the axon per unit length is $R_i$, so the series resistance between two points separated by a small distance d$x$ along the core of the axon totals $R_i$d$x$ as explained above. The cell membrane is assumed to have a capacitance of $C_m$ per unit length and a resistance of $R_m$ per unit length, so the magnitude of these quantities for the capacitor and for the resistor connecting the axoplasm to the external fluid over the length d$x$ are $C_m$d$x$ and $R_m$/d$x$ respectively.

What are the implications of this in practice? The resistivity of axoplasm is about 0.6 ohm m, the membrane resistance is about 0.2 ohm m$^{-2}$, and the capacitance of a unmyelinated axonal membrane is about $10^{-2}$ F m$^{-2}$. For the squid axon which has a diameter, $d$, of 1 mm, the calculated values of $R_i$, $R_m$ and $C_m$ are:

$$R_i = 7.64 \times 10^5 \text{ ohm m}^{-1}$$

$$R_m = 63.7 \text{ ohm m}^{-1}$$

$$C_m = 31.4 \,\mu\text{F m}^{-1}$$

▷ A more typical diameter for an animal axon is 50 µm. Assuming it is unmyelinated, make an estimate of $R_i$, $R_m$ and $C_m$ for such an axon by using the values calculated for the squid axon with a diameter of 1 mm.

▶ The diameter is 20 times smaller than that of the squid axon. Thus, we would expect

$$R_i = 7.6 \times 10^5 \times 20 \times 20 = 3.04 \times 10^8 \text{ ohm m}^{-1}$$

$$R_m = 63.7 \times 20 = 1.274 \times 10^3 \text{ ohm m}^{-1}$$

$$C_m = 31.4/20 = 1.57 \,\mu\text{F m}^{-1}$$

So, for a 100 mm length of the 50 µm diameter axon, the resistance along the axoplasm is 2 385 times bigger than the leakage resistance across the membrane, while for the same length of the 1 mm diameter squid axon it is 119 times bigger. Therefore, most of the charge generated does not pass along the axon (the cable) as a signal, but is lost through the the membrane (the insulation). Thus the axon is a very poor cable indeed compared with a domestic copper-cored insulated cable – in a 100 mm length of the latter, the resistance along the core is entirely negligible in comparison with the leakage resistance through the insulation. Not surprisingly, it has been said that the axon has all the desirable cable characteristics of a piece of wet string! This is one reason why constant amplification of the signal is necessary as it is propagated.

Now, if a capacitor with capacitance $C$ is charged to an initial voltage $V_0$, and then allowed to discharge through a resistance $R$, the voltage $V(t)$ will decay with time $t$ during the discharge as

$$V(t) = V_0 \exp(-t/RC)$$

When $t = RC$,

$$V(t) = V_0 \exp(-1)$$

$$= V_0/\exp(1)$$

$$= V_0/2.72$$

We can use this formula to estimate the time it takes for the positive charge in the axoplasm to leak away through the membrane which has resistance $R_m$. If a unit length of an axon is charged to a membrane voltage $V_m = V_0$, then this voltage will decay to $V_m = V_0/2.72$ in a time $t_m$, given by $R_m C_m$. This product is $2 \times 10^{-3}$ s if the values of $R_m$ and $C_m$ for the 1 mm diameter squid axon are used. Note that, because $R_m$ is inversely proportional to the membrane area and $C_m$ is directly proportional to that area, the product $R_m C_m$ does not depend on the diameter of the axon.

The speed with which the action potential propagates depends not only on the passive cable properties that we have discussed, but also on the rate of change of the membrane voltage when the channels open, and this in turn depends upon the conductance of the channels, their density in the membrane and on many other factors. With so many variables involved it would be no simple task to determine the speed precisely, but computer models can mimic the behaviour of an action potential with good accuracy by using equivalent circuits for the axon (like that in Figure 5.6b) in conjunction with models of ion-channel action.

Among the various other factors that can influence the speed of propagation of the action potential is the degree of myelination, as has been mentioned previously. We shall now consider how this factor exerts an influence. In myelinated axons the ion channels are located only in the nodes of Ranvier between the myelinated sections. The signal is amplified by the ion channels in these nodes, but between the nodes the signal is propagated by passive ionic diffusion along the lengths of axoplasm insulated by the myelin sheath. The myelin sheath can result in the membrane resistance per unit area increasing by a factor of 100, and the membrane capacitance per unit area decreasing by approximately the same factor. This lowers the time constant, $t_m = R_m C_m$, for charging the interior of the axon, and thus speeds signal propagation along the myelinated sections. Moreover, lowering the capacitance means that less charge has to be conveyed to raise the voltage, and increasing the membrane resistance means that less charge leaks away – both factors enhancing the efficiency of the signal propagation. For axons with a diameter greater than about 1 μm, a myelinated axon will conduct at a greater speed than a unmyelinated axon of the same diameter. Indeed, calculations show that the signal speed is approximately proportional to the square root of the diameter for unmyelinated axons, and approximately proportional to the diameter in myelinated axons. Experiments tend to support these predictions. For axons with a diameter of less than 1 μm, theory predicts that myelination confers no speed advantage.

In conclusion, the propagation speeds of $1\,m^{-1}s$ to $100\,m\,s^{-1}$ that occur in long nerve axons are many orders of magnitude greater than is possible by diffusion over these distances. They are also much faster than is possible in the more basic cell chain mechanism described in Section 5.2, but they are still very slow in comparison with electron currents in metal and semiconductor communications systems. To send an action potential from one side of the head to the other takes about 1 ms, but in that time a modern computer can complete tens of thousands of arithmetic calculations. However, in nerve signalling, the lack of speed is compensated for by extensive parallel operation with many signals being processed simultaneously – a system that is only just being introduced into computer architecture.

### Summary of Section 5.3

Electrical signals may be sent rapidly by neurons between particular locations that are widely separated. A region of the cytoplasm is made to carry a positive charge, and this region propagates along the axon of the excited neuron. The synchronous opening and shutting of voltage-gated ion channels across the membrane of the axon in the region of the positively charged axoplasm enhances and amplifies the action potential. Amplification along the length of the axon is necessary to compensate for signal lost by leakage across the axon membrane.

## Objectives for Chapter 5

When you have completed this chapter, you should be able to:

5.1   Define and use, or recognize definitions and applications of, each of the terms printed in **bold** in the text.

5.2   Describe how an action potential may be propagated along a chain of cells that are connected by gap junctions, and state which properties of the cells determine the speed of propagation.

5.3   Describe how an action potential is conducted along an axon.

5.4   Describe in what respects the axon of a neuron resembles an electrical cable, and explain the need for constant amplification of the signal if the signal is not to decay rapidly in both space and time.

5.5   Distinguish between the mechanisms of signal propagation in myelinated and non-myelinated axons.

5.6   List the advantages, compared with simple diffusion, of the propagation of action potentials along chains of connected cells or along nerve axons when electrical signals travel between specific widely separated cells.

## Questions for Chapter 5

### Question 5.1  *(Objective 5.3)*

Why is the speed of propagation of an action potential along a chain of cells of a given diameter usually much slower than the speed of propagation of an action potential along an axon of the same diameter ?

*Question 5.2 (Objective 5.2)*

Given that the capacitance, $C$, of a living cell is proportional to its surface area, how would you expect the speed of propagation of an action potential along a chain of cells to vary with the cell diameter? Assume that the characteristics and total resistance, $R$, of the gap junctions that couple the cells remain constant.

*Question 5.3 (Objective 5.3)*

What would be the effect on the passage of the action potential along an axon if a poison were applied that had the effect of (a) blocking some of the $Na^+$ channels *or* (b) blocking some of the $K^+$ channels?

*Question 5.4 (Objective 5.6)*

Discuss the reasons why ionic diffusion may be suitable for the transmission of electrical signals across a single cell but is not suitable for communication between cells that are separated by greater distances.

## CHAPTER 6

**Living cells and electric and magnetic fields**

## 6.1 Introduction

Having established the biophysical principles of electrical signalling processes within a single resting cell, and then within an entire organism via its nervous system, we now turn to a rather less conventional theme – the relationship between living cells and electric and magnetic fields – a theme which builds upon the understanding of bioelectrical processes that you should have acquired from Chapters 2 and 5.

Voltage differences and current flows exist across all biological membranes, and, as you have just seen in Chapter 5, action potentials are transmitted along excitable membranes. All this electrical activity means that living cells and tissues create electric fields and electrical currents around themselves and, hence, magnetic fields also. Biologists have been slow to recognize the importance of such fields for the understanding of living processes. Indeed, the implications are still a matter of debate as, for example, in the case of the brain. Brain processes are usually described in terms of the specific conduction of nerve impulses and synaptic signalling, but the associated myriad of action potentials gives rise to biomagnetic and bioelectric fields: such fields may be an epiphenomenon of the activity of neurons, or they may have informational significance for the organism. Rival theorists still contest these possibilities.

It is not only such internally produced electric and magnetic fields that are significant for living systems: external fields may have important influences too. For instance, many species have evolved special methods of detecting and utilizing external electric and/or magnetic fields – perhaps the best known examples are the exploitation of the Earth's magnetic field by migratory birds as an aid to navigation, and the detection of electrical signals by deep-sea fish to detect prey or to navigate. Other examples exist and are now slowly being identified.

The potential sensitivity of living systems to external electric and magnetic fields has raised some concerns, since many human technologies (from microwave ovens and computer display screens to overhead power lines) generate electric and/or magnetic fields around the respective devices. The effect on living systems – including, of course, humans – of such constant exposure to such low-intensity radiation is a matter of interest and investigation. However, external fields should not only be seen as a potential threat; in fact, it has been possible to harness externally generated fields to elucidate the structure of living tissues. For example, nuclear magnetic resonance imaging, which involves exposing living organisms to large magnetic fields, has become an important analytical tool, particularly for clinical diagnosis.

This is a fast-developing area, attracting much research effort as well as media coverage – so it is interesting and timely to consider it in more depth. Here we shall discuss the detection of the electric and magnetic fields created by living cells (Section 6.2); we shall then look at an example of the controlled application of an externally generated field and go on to explore the influence of externally generated ambient electric and magnetic fields upon living cells (Section 6.3).

## 6.2 Detection and significance of electric and magnetic fields produced by living cells

The microscopic currents generated by the electrical activity of individual cells may combine to produce a current that can be detected non-invasively from outside the organism. Such currents are studied for a variety of reasons ranging from fundamental investigations into life processes to the diagnosis of disease. The biologically generated currents are very small and, in many cases, it is only by using the most advanced instrumentation that they can be detected and analysed, either directly, by measuring the current itself, or indirectly, by measuring the electric or magnetic field due to the current.

We shall illustrate such measurements by an investigation of the currents generated by a developing embryo (Section 6.2.1) and by a look at studies on the functioning of the human brain carried out by means of sensitive superconducting quantum interference device (**SQUID**) magnetometers (Section 6.2.2).

### 6.2.1 Breaking symmetry in the developing embryo: the vibrating probe electrometer

In the initial stages of cell division, an embryo tends to be highly symmetrical. There are no obvious directions – no left or right, no head or tail, no up or down. However, at some stage the cells must differentiate direction so that particular regions of the embryo develop in individual ways. Some internal or external interaction must *break the symmetry:* among the candidates for such an interaction are gravity, neighbouring spatial interactions, local electric fields, local ionic or chemical gradients, or even the direction of incident light.

Let us consider this in more detail in the case of the embryos of brown algae such as *Pelvetia*, which are much studied. Initially, the fertilized egg appears to have spherical symmetry, but later in its development a rhizoid outgrowth develops at a particular point on the embryo's surface and an axis of symmetry is established (Figure 6.1).

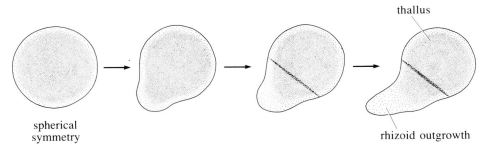

thallus

spherical symmetry

rhizoid outgrowth

**Figure 6.1** Symmetry breaking in the developing Pelvetia embryo.

The development of the rhizoid is known to involve the local secretion of wall-softening enzymes and the pumping of $K^+$ and $Cl^-$ ions into the cell. The position of the outgrowth can be influenced by illumination from a particular direction or by local chemical gradients.

So what might be going on? As you know from Chapter 2, all living cells establish ionic concentration and voltage gradients across their outer membranes. In the steady state, the efflux or influx of a particular type of ion is balanced over the entire cell by the flow of that ion through ion channels in the membrane and by membrane leakage. But if the balance is not maintained locally, and regions of the cell surface experience a net influx of a given ion while other regions have a net efflux, then a current due to that ion will be established across the cell. Ions expelled from some regions of the cell membrane and absorbed over other regions also create ion currents which flow in the external fluid bathing the cell. The internal current across the cell implies either a concentration gradient for the given ion or a voltage gradient across the cell or possibly both. Now, in the case of *Pelvetia*, as early as 30 minutes after fertilization of the egg, the external ionic currents can be detected flowing around the embryo, and their role in the development of the rhizoid is of considerable interest.

The ionic currents may be measured in the fluid bathing the embryonic cells by means of a *vibrating probe electrometer*. This electrometer consists of a sensitive voltmeter connected through an insulated cable to a small metal tip which is made to vibrate linearly at a fixed frequency through a small fixed distance. If an electric field exists in the fluid around the tip of the probe, the maximum voltage difference experienced by the tip as it vibrates will depend on the distance through which the tip vibrates (typically 30 μm) and on the component of the electric field along the direction of the tip's vibration. Hence that component can be determined and, if the electrical conductivity of the fluid is known, the current density in the given direction can then be calculated. The currents generated are very small (much less than 1 μA) and the resultant voltage change over a tip vibration is only about $10^{-8}$ V, so good sensitivity is required. In practice, the detection of the signal may be sharply tuned to the vibration frequency, and the signal-to-noise ratio may be as high as 1 000 times that of a static voltage probe.

The initial current pattern detected for *Pelvetia* embryos is unstable with more than one region of net influx, but later the pattern stabilizes and becomes fixed with the current entry point accurately predicting the region where the rhizoid will form some eight hours later. A positive electric current flowing in a given direction may be due to cations flowing in that direction or to anions flowing in exactly the opposite direction. Experiments show that the inward current is carried mostly by a $Cl^-$ ion efflux but also partly by a $Ca^{2+}$ ion influx. The $Ca^{2+}$ ion current that traverses the embryo is about $2 \times 10^{-12}$ A. It is thought that the $Ca^{2+}$ ion gradient has the central role in breaking the original symmetry and determining the polar axis (see Figure 6.2). This tends to be borne out by the finding that manipulation of the direction of the internal $Ca^{2+}$ ion gradient changes the orientation of the subsequent axis of polarization of the embryo, and that inhibition of this gradient prevents polarization altogether.

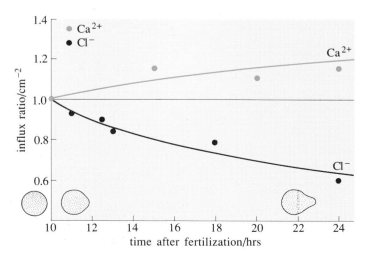

**Figure 6.2** Calcium and chloride ion gradients and symmetry breaking.

As you know from Chapters 1, 3 and 4, $Ca^{2+}$ ions are employed as messengers in many living systems, because the free $Ca^{2+}$ level is maintained at very low concentrations (in the micromolar range) within the cells and, consequently, the influx of even a small number of $Ca^{2+}$ ions represents a large perturbation, producing an easily detected signal. It is known that $Ca^{2+}$ gradients and $Ca^{2+}$ ion flows occur across the embryos of many species – in some cases the flow is steady and in others it fluctuates with time to form waves. There seems to be little doubt that such coherent currents are often key indicators of the reduction in symmetry associated with the next stage of embryo development. In other cases of symmetry-breaking in developing embryos, even in other algae, the situation is less clear cut, and it is possible that several symmetry-breaking forces may be at work.

### 6.2.2 Magnetometer determination of currents in the brain: the superconducting quantum interference detector

In humans, the measurement of voltages detected by electrodes attached to the skin has played a large part in the diagnosis of aberrant behaviour of the heart or the brain. Although important clinically, the signals detected are only indirectly related to the electrical activity of the underlying excitable tissue, because of the presence of intervening muscle, bone and skin. Recently, more direct information about the electrical activity of excitable tissues has been obtained by using a very sensitive detector of magnetic fields called a **Superconducting Quantum Interference Detector** or **SQUID** (see Figure 6.3). Magnetic fields are created by electrical currents, and by measuring the magnetic field around a cell it is often possible to deduce information about the ionic currents flowing in and around the cell.

A SQUID can reliably detect magnetic fields as low as $10^{-14}$ T (T denoting tesla – a unit of magnetic field strength), provided that magnetic interference from external sources can be reduced sufficiently and that the specimen is free from

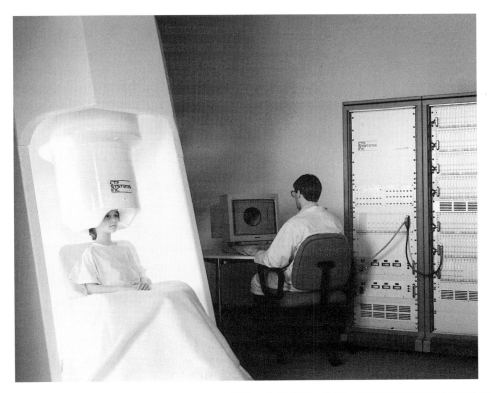

**Figure 6.3** The components of a multi-channel SQUID magnetometer system. The patient sits with his/her head inside a helmet containing the sensors. In routine clinical operations the system, which is controlled by large computers, would be inside a magnetically screened room.

magnetic impurities; specially built SQUIDs have even been used to determine the current produced by an action potential passing along a single axon. (To refresh your memory about the size of magnetic fields, a powerful iron-cored electromagnet may generate about 1 T, while the Earth's magnetic field is about $50 \times 10^{-6}$ T.)

There are some caveats about the use of SQUIDs. Often, the magnetic fields due to the currents generated by living systems are a million times smaller than the stray magnetic fields found in a typical laboratory as a result of the electrical currents generated by, for example, lighting, heating and electric motors. This means that, ideally, an experiment with a SQUID must be performed in a magnetically screened room so as to prevent interference from externally generated magnetic fields. Another problem arises in the interpretation of the data obtained from SQUIDs (and other magnetometers); it is the so-called 'inverse problem'. It occurs because, despite the fact that a given current distribution produces a unique magnetic field, the inverse is not true in that a given magnetic field pattern may be produced by many different current distributions. Thus, to obtain information about currents from a knowledge of the magnetic fields they produce requires that certain assumptions are made about the geometry of the current sources. For example, when trying to deduce the currents flowing in an active brain by studying the magnetic field these currents generate, one may have to assume that the current is generated from a small number of localized sources or is confined to a superficial shell.

However, by using large arrays of SQUIDs close to the human head and relying on the speed of modern computers to interpret the measurements, much progress has been made recently in the development of SQUID magnetometers for the investigation of brain behaviour. These allow the determination of the distribution of currents, and further processing enables that distribution to be superimposed on an image of the brain structure (Figure 6.4).

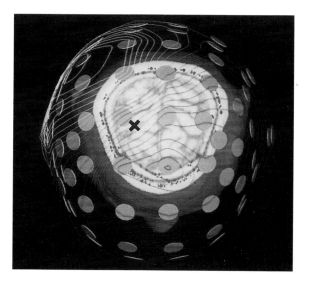

**Figure 6.4**  The brain's response (highly localized) to stimulation of a nerve in the leg. The site of the activity, shown as a black cross, is in the somatosensory cortex. It is superimposed on an MRI image of the brain (see later) and a map of the SQUID sensor locations.

As yet, the spatial resolution with which the currents in the brain can be localized is only about 10 mm (unless the system is particularly simple). As you will see in the next section, this does not compare favourably with other techniques that can be used to elucidate the brain's structure. Yet the time resolution is very good: changes in the currents over time periods of the order of one millisecond can be detected, which is faster than any other known technique for the non-invasive detection of currents in the brain.

To date, one of the most productive applications of SQUID technology has been to measure the response of the brain to a repetitive light or sound stimulus. These experiments are capable of a high signal-to-noise ratio as only the magnetic fields that are synchronous with the stimulus are averaged, while fields resulting from other brain activity tend to average toward zero. Consider an experiment in which a brief flash of light is generated at irregular intervals and during which the measured magnetic fields around the brain are recorded with each record starting only when the light flashes. The brain response evoked by the flashes will have the same time dependence in each record so that as an increasing number of records are added the signal evoked will become progressively bigger. Other brain activity, not connected with this light flashing, will not have the same time dependence in each record and hence the (measured) magnetic fields will tend to cancel out when the records are added. Thus, by adding many records synchronized by the light flash, the required signal may be selectively amplified to dominate over all the others. In this way, information about the brain's processing of such stimuli is gained, and differences between normal and aberrant responses may, in principle, be interpreted clinically, though such technologies are not yet in routine use.

## 6.3   Interaction of living cells with external fields

In the preceding section we have considered some aspects of the electric and magnetic fields that are generated internally by living cells. But it is not only these *internal* fields that are important in living processes, *external* fields can interact with living cells in various ways. The interaction depends on the nature and, particularly, the frequency of the field: very high-frequency, and thus high-energy, electromagnetic radiation such as gamma-rays, X-rays and ultraviolet light damage living cells by colliding with biologically important molecules in the cells and so directly causing damage. At lower radio-frequencies, electromagnetic radiation of sufficient amplitude may cause damage when it is absorbed by biological molecules and gives rise to local heating as in a microwave oven. Here we are not concerned with such ballistic and thermal damage, but with the effects of applying electric and magnetic fields of much lower frequency and energy. We shall first look at a controlled application of an external field, specifically an external magnetic field used in conjunction with electromagnetic radiation in a medical diagnostic technique based on **nuclear magnetic resonance (NMR)**. We shall then go on to enquire into how living systems can be affected by externally generated low-frequency electric and magnetic fields that occur in the general environment and are relatively weak, i.e. of small amplitude – sometimes to the system's advantage and sometimes to its possible detriment.

### 6.3.1   Imaging the brain: nuclear magnetic resonance measurements

In living systems many molecules, from water to phospholipids, contain hydrogen atoms. The nucleus of a hydrogen atom consists of just one proton. In a magnetic field, this nucleus has two magnetic energy states arising from the interaction of the magnetic moment of the nucleus with the magnetic field. The energy difference or splitting between the two states or levels, $U_{\mathrm{mag}}$, is proportional to the magnetic field, $B$, so

$$U_{\mathrm{mag}} = K_{\mathrm{H}}B \tag{6.1}$$

where $K_{\mathrm{H}}$ is a constant, characteristic of the hydrogen nucleus.

Now, if a sample containing molecules that incorporate hydrogen atoms is placed in a constant magnetic field, $B$, and a radio-frequency electromagnetic field of frequency $v$ is applied, the nucleus of a hydrogen atom will absorb energy from the radio-frequency field only if the energy $hv$ of a photon of the electromagnetic radiation corresponds to the energy difference between the two split nuclear magnetic energy levels in the atom, i.e. when there is resonance and

$$hv = K_H B \tag{6.2}$$

But the field experienced by a given hydrogen atom in the sample will, in fact, be $B + \Delta B$, where $\Delta B$ is a small change in the local field due to the magnetic moments induced by the applied field, $B$, in molecules within the sample that are close to the given hydrogen atom. The resonant frequency of that hydrogen atom will be given by $(v + \Delta v)$ in the following equation, which is analogous to Equation 6.2:

$$h(v + \Delta v) = K_H(B + \Delta B) \tag{6.3}$$

By studying the distribution of values of $\Delta v$ for the hydrogen atoms within the sample, one can obtain information about their environments. For example, the hydrogen atoms in a $CH_2$ group will have a different value of $\Delta v$ from those in a $CH_3$ group. In this manner, the internal structure of the sample may be determined. For technical reasons, it is more convenient to keep the applied frequency, $v$, fixed and sweep the magnetic field, and then record the strength of the detected resonance of the hydrogen atoms in the sample as the applied field is varied. The principle remains the same in that hydrogen atoms in particular molecular environments will resonate at particular measured and characteristic applied fields. This is the basis of the technique of nuclear magnetic resonance. The technique may also be employed to detect nuclei other than those of hydrogen, but the sensitivity is much less than for the detection of hydrogen.

The technique is normally used as an analytical tool to determine the atomic constitution of a sample as the overall response reflects the overall constitution and structure of the system. As described in Book 1 (Video sequence 1), the complete three-dimensional structure of a small protein molecule in solution may be deduced by these means, but here we are interested in another application of NMR, that is **Magnetic Resonance Imaging (MRI)** – a non-invasive technique that can be used to map the internal structure of a living organism.

Consider a biological specimen placed in an essentially uniform applied magnetic field that has a small field gradient superimposed on it in the $x$-direction. For simplicity we shall ignore the small changes in the local field experienced by a particular hydrogen atom as discussed above. The hydrogen nuclei in each plane, of thickness d$x$, perpendicular to the $x$-direction will experience a slightly different total magnetic field because of the gradient; it follows from Equation 6.1 that the difference, or splitting, between the magnetic energy levels of a nucleus will depend on the plane d$x$ in which that nucleus is located. When a radio-frequency field is applied, only those nuclei in the particular plane to which the magnetic field is attuned will absorb energy from the applied field. Thus the NMR absorption that is measured is proportional to the number of hydrogen nuclei within the particular plane in the specimen. The measurement may be repeated many times at slightly different uniform applied magnetic fields so that the density of hydrogen nuclei is determined within a different plane or 'slice' each time; in this way the distribution of hydrogen nuclei may be obtained as a function of $x$ throughout the specimen. If the whole sequence is repeated with magnetic field gradients applied along the $y$- and $z$-directions in turn, a complete three-dimensional map of the hydrogen atom distribution in the whole specimen may be obtained.

As such a large part of the living cell is water, cells have a high hydrogen-atom content, so a good high-resolution image of a specimen may be obtained, as Figure 6.5 shows. There is even considerable contrast between different aqueous regions because of differences in the rate at which the irradiated hydrogen nuclei can dissipate the energy absorbed from the irradiation. In fact, the spatial resolution achieved to date in studies of humans is about 1 mm – sufficient to enable abnormalities, such as the sites of brain damage due to stroke, tumour or atrophy, to be detected non-invasively.

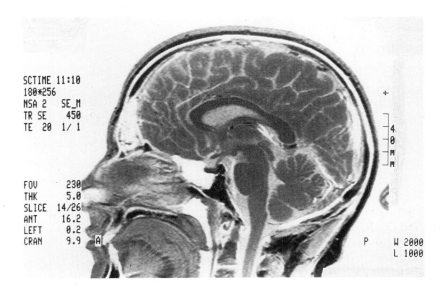

***Figure 6.5*** MRI plot of a slice through the the brain of a human subject. The darkest parts of the image correspond to regions of high proton density.

Modern MRI systems make use of varying fields and radio-frequency pulses which extract information from the system very efficiently. When combined with advanced computers to interpret and store the data, they are capable of producing a complete three-dimensional hydrogen-atom image of a human heart or brain in a time as short as 30 ms. The distribution of hydrogen atoms within a slice of the specimen, taken at any orientation or position, may later be recalled from the computer memory and displayed on a screen. This is the basis of an important technique called **tomography**, whereby a specimen is viewed slice by slice; it is much used for medical diagnosis.

But how can the image of, say, the brain produced by MRI be related to the electrical activity within the organ? The key fact exploited to relate structure and activity is that the intensity of the image depends to a certain extent upon the state of oxygenation of the blood. Oxygen is carried in the blood by haemoglobin which becomes slightly magnetized when placed in a magnetic field, i.e. paramagnetic, in its non-oxygenated state. The small local fluctuating magnetic fields created by these paramagnetic molecules change the NMR signal. By focusing on this difference, it has proved possible to produce a three-dimensional image which emphasizes regions of the specimen with high proportions of deoxygenated blood. As both the passage of an action potential and transmission across the synapse require energy, it follows that neural activity depletes the oxygen content of blood and so produces more deoxygenated blood: the MRI image will therefore highlight regions of unusually high activity. Although it is not the electric currents in the cells themselves that are imaged but the blood supply to the nerves that generate the currents, the spatial resolution of the activity maps is sufficient to make possible the study of many aspects of the working of the brain or other organs such as the heart. However, the time resolution of the technique is inferior to that provided by SQUID methods because changes in the blood flow lag behind the neural activity and are smeared out in time. For this reason, the technique is mainly limited to changes that take a large fraction of a second to occur.

A great strength of the MRI technique is that it automatically records and superimposes the neural activity map onto the accurate structural map so current flows may be correlated with known features. Thus, an accurate three-dimensional record of the topography of the heart can have superimposed upon it the major nerve conduction paths with an indication of their amplitude over time. In another example of the technique, the spatial patterns of nervous activity throughout the brain when the subject is looking at an object may be compared with the brain activity evoked when the subject merely recalls the same object with closed eyes. The procedure is non-invasive and quite rapid. As a result, it appears to be a very powerful new tool for increasing understanding of large-scale electrical activity in living animals.

Tomography can be based on other nuclei besides hydrogen, for example phosphorus. Thus energy storage and utilization may be studied by measuring the relative amounts and locations of ATP, ADP and inorganic phosphate. However, the much lower sensitivity limits the usefulness of the technique for measuring such metabolites.

### 6.3.2   Interaction with ambient static magnetic fields

Many animals migrate over vast distances every year. Others move in an apparently featureless environment, like a lake or ocean, but need to be able to return regularly to known habitats. In either case, some sense of direction is required. It has been established that such animals use a whole spectrum of navigational techniques, including interpreting star maps or polarized light maps and even following scent trails. It is also now well established that many creatures can detect the direction and strength of the Earth's magnetic field; among those that use such magnetic information are flies, newts, fish and birds. The best understood example of magnetic field detection is provided by certain sediment-dwelling bacteria. These magnetotactic bacteria (see Figure 6.6) move along the direction of the Earth's magnetic field. It seems that knowledge of the direction of the Earth's field may be important to them in the mud or sea as an indication of 'up' or 'down'.

On examination, each bacterium is found to contain a row of single crystals of **magnetite**, a ferrimagnetic iron oxide with the chemical formula $Fe_3O_4$. Magnetite crystals form strong magnets if their major dimension is between about $0.01\,\mu m$ and $0.5\,\mu m$. Crystals that are smaller than this have no permanent magnetic moment, and an applied magnetic field induces only a weak moment. Crystals of magnetite that are larger in any dimension than about $0.5\,\mu m$ will contain oppositely magnetized domains, so their total permanent magnetic moment is small. The magnetite crystals within magnetotactic bacteria have their size controlled by the organism to be in the range of strong magnets (i.e. $0.01$–$0.5\,\mu m$); moreover, the magnetic moments of all the crystals in the bacterium are aligned. The mechanism whereby such a simple organism can control the size and alignment of the crystals is not fully understood, but it is known that the crystals are confined within membranes called **magnetosomes** which only allow growth

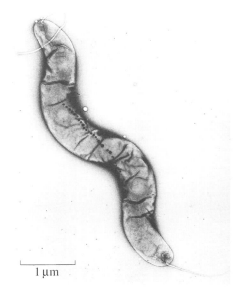

$1\,\mu m$

**Figure 6.6**   A magnetotactic bacterium.

in certain directions. Since the original discovery of the bacteria, other magnetotactic bacteria have been found which live in sulphur-rich environments and use ferrimagnetic iron sulphide crystals rather than iron oxide.

How can the the bacterium utilize these magnetic crystals to acquire directional knowledge? A magnet of magnetic moment $M$ experiences no translational force in a uniform magnetic field $B$, but it is acted on by a couple of strength $MB \sin \theta$ where $\theta$ is the angle between the moment and the field. This couple tends to rotate the magnet and thus align the magnetic moment along the magnetic field. Magnetite forms such a powerful magnet that a single crystal of the size commonly found in bacteria, i.e. with dimensions $0.1 \, \mu m \times 0.05 \, \mu m \times 0.05 \, \mu m$, will, if it is free to rotate, be more than 4.5 times as likely to be pointing along the Earth's field as against it. This is true despite the thermal buffeting that the crystal will receive in a fluid at room temperature and despite the very weak amplitude of the Earth's field (about $50 \times 10^{-6} \, T$).

Magnetotactic bacteria are forced to swim along the direction of the Earth's field, by the strength of the magnetic couple acting on the magnetic crystals; in the Northern Hemisphere they are oriented towards the North Pole, i.e. with the north-seeking (north) pole of each crystal and, therefore, of the row of magnetic crystals overall to the fore. The aligned motion of the bacteria when a magnetic field is applied may be observed under a microscope. If a short pulse of a very high magnetic field is applied, sufficient to reverse the direction of the magnetic moments of the magnetite crystals, the bacteria are observed to reverse their direction of motion accordingly.

Many larger animals have been shown to use the direction of the Earth's magnetic field as a navigational aid. In the case of migrating birds, it has been shown that a bird's use of the magnetic field information is more complicated than simply utilizing a single compass. Figure 6.7 illustrates the three aspects to be considered: (i) the Earth's ambient magnetic field, (ii) the conventional needle compass, and (iii) the flight direction of a bird attempting to fly southwards from a starting point in the Northern Hemisphere. The ambient magnetic field, $B$, can be resolved into a vertical component, $B_V$, and a horizontal component, $B_H$, (see Figure 6.7). Reversing the *vertical* component of the ambient magnetic field leaves unchanged the direction in which the conventional compass needle points (Figure 6.7b), but reversing the *horizontal* component reverses the direction of the needle, as indicated in Figure 6.7c. In contrast, experiments have revealed that the flight direction of the migrating bird is reversed if *either* the vertical *or* the horizontal

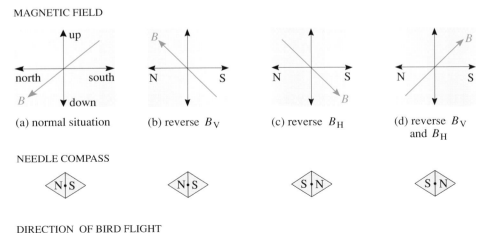

Figure 6.7   The magnetic compass used by birds.

component of the ambient magnetic field is reversed. However, a complete reversal of the magnetic field direction (see Figure 6.7d), which reverses both the vertical and horizontal components (and which reverses the direction of the conventional compass), is not detected by the bird. It appears that the bird can detect the direction of both the magnetic field and of gravity. North is then interpreted to be the direction in which the magnetic field axis and the downward direction of gravity make the smallest angle. This is the basis of a device called an **inclination compass**. Some experiments suggest that a bird may even be able to sense the acute ($< 90°$) angle between the direction of the magnetic field axis and the downward direction of gravity with sufficient accuracy for it to serve as a measure of latitude, since this angle is $0°$ at the magnetic pole and $90°$ at the magnetic equator.

Magnetite crystals have been found in the tissues of many animals, including in the heads of all humans and in those of some migrating birds, but it is not yet established whether the magnetite is used to detect the Earth's field. It may simply represent a store of iron, although it is difficult to see why, if that is the case, it is always found to have the crystal size necessary to form single-domain magnets.

Another type of magnetic field detector is known to be present in many animals: it utilizes light, with the eye as a sensor. Examples are found in the fly, the newt and in some birds, such as pigeons. In the head of a pigeon, certain nerves – some of which are thought specifically to originate in the eye via the optic nerve – have a rate of firing that depends directly on the magnetic field in which they are placed. Some of these nerves respond to a change in the *amplitude* of the field, while others respond to a small change in *direction* of an applied field of constant amplitude. In fact, there is evidence of at least three distinct types of field detector in the pigeon, one of which depends on light reaching the eye. Other evidence for an optically-based magnetic field detector lies in the effect of changing the wavelength of the light incident upon the animal. In these cases an animal trained to indicate the field direction while exposed to a particular wavelength of light can perform well in this light but will make large errors in orientation, often of $90°$, when performing in light of a different wavelength. Some migrating birds trained in blue or green light respond normally in these lights but entirely lose their ability to detect the Earth's field in red light. Although laboratory techniques do exist for the precise optical detection of the direction and amplitude of small magnetic fields, no entirely satisfactory biological model to test these intriguing findings in animals and birds has as yet been constructed.

### 6.3.3  Interaction with externally generated electric fields

All cells rely for their function on mobile ions, and these interact strongly with electric fields. Moving ions also interact with magnetic fields: such a field exerts a force on a moving ion, but *not* on a stationary ion. If we compare the electric force $F_e$ and the magnetic force $F_m$ which act upon a given charged particle, such as an ion, when it moves through *static* (i.e. steady or unvarying) electric or magnetic fields of the same energy density, then

$$F_m/F_e = v/c \qquad (6.4)$$

where $v$ is the velocity of the particle, and $c$ is the velocity of light. Thus for slow-moving charged particles, like ions diffusing in an aqueous fluid, the magnetic forces are entirely negligible in comparison with the electric forces provided that the electric and magnetic fields have the same energy density.

Clearly, this does not comprise a complete analysis of the interaction of external fields with complex systems containing not just free ions but other more complicated molecules and structures. However, it does suggest that living cells would be sensitive to electric fields but would not be affected by magnetic fields – but, as we have just seen, this conclusion would be wrong. Some animals utilize a knowledge of the Earth's magnetic field as a navigational aid and have evolved special magnetic field detectors. On the other hand, it is interesting that the conducting tissues of living animals are very effective in excluding externally generated electric fields from their interior.

Because the cell cytoplasm and the extracellular fluid both conduct electricity, but the resting membrane does not, any low-frequency externally applied electric field is rapidly excluded from the cell interior. The underlying principle here is that no electric field may exist within a conductor in equilibrium: if an electric field is applied, then free charges within the conductor move under the influence of the field until the field is cancelled to zero. To illustrate this, consider a parallel-sided 'slab' of conducting fluid held between non-conducting 'sheets' which occupy the planes $x = 0$ and $x = A$, as shown in cross-section in Figure 6.8. An external electric field, $E_e$ , is applied in the direction of positive $x$. Under the influence of the field, some mobile positive charges in the conducting fluid will move in the positive direction of $x$ until they reach the non-conducting sheet (plane $x = A$), while negative charges will move similarly but in the opposite direction. The net result will be a sheet of accumulated positive charges in the plane $x = A$ and a sheet of accumulated negative charges in the plane $x = 0$. These sheets of charge will produce a uniform internal electric field, $E_i$, in the negative direction of $x$, and this will tend to cancel the original field, $E_e$. So long as *any* electric field remains within the interior of the conducting fluid, charges will tend to move so as to counter that field, continuing to do so until the original field is exactly reduced to zero when the amplitude of $E_e$ equals that of $E_i$.

This general argument holds for the interior of a conductor of any shape or size except in the immediate vicinity of its surface where the mobile charges that screen the applied field accumulate. Thus, if an external electric field were applied to a dish containing isolated living cells (a tissue culture) immersed in a medium with a composition resembling that of extracellular fluid, very little of the field would penetrate the fluid surrounding the cells, owing to the accumulation of mobile charges at the interface between the dish and the conducting fluid. Any residual fields that remain in the bathing fluid would be effectively excluded from the cell cytoplasm by other charges accumulating at the inner surface of the cell membrane. Very near the non-conducting barriers (such as the dish and the cell membranes) where the mobile charges accumulate, the electric field might even be greater than the original applied field.

The time taken to cancel the applied field in the bulk of a conducting fluid depends on the dielectric constant and the electrical conductivity of the conducting fluid. With typical values of these quantities for the cytoplasm, the calculated time to cancel the field is very short – of the order of $10^{-9}$ s. So, an external electric field will penetrate the cytoplasm only for a very short time before it is cancelled by the fields set up by mobile charges.

Recent epidemiological evidence that the incidence of some childhood leukemias and brain tumours may be higher in those living close to high-voltage overhead power lines has stimulated renewed interest in the effects of electric (and magnetic) fields on living cells. Under such a power line, the electric field is sinusoidal with a frequency of 50 Hz and an amplitude possibly

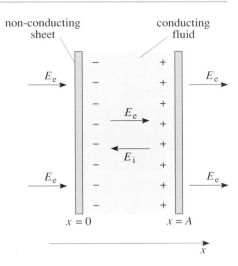

**Figure 6.8** Electric fields across a conducting fluid held between sheets of a non-conductor.

as high as $10\,kV\,m^{-1}$. Because a 50 Hz field changes very little in the time required to screen the field from a cell, it is, like a static electric field, strongly excluded from any conducting body.

Because of the exclusion of the electric field from the body of the cell and the accumulation of mobile charges near non-conducting membranes where the electric field may be enhanced, much attention has been focused on the cell membrane as the site for the interaction. However, the field across the cell membrane of a typical resting animal cell is very large – of the order of $10^7\,V\,m^{-1}$ – even in the absence of an externally applied field. To have an effect, the external field must appreciably modify the resting field. In view of the large attenuation of the $10\,kV\,m^{-1}$ field (generated by the power line) by the intracellular and extracellular conducting fluids, it is difficult to imagine how that external field can appreciably modify the effects produced by the very large natural internal electric field. This conclusion would be substantially affected by the enhancement of any applied field near the membrane because of the accumulated mobile charges.

The result of such calculations has been to divert the search for any causal effect of the field towards some possible trigger process or towards a mechanism with a large amplification such that a small stimulus can give rise to a much larger effect.

### 6.3.4 Electric fields induced by varying magnetic fields

Electric fields may be generated by static arrays of charges, as in the case of the internal field $E_i$ in Figure 6.8. However, such fields can also be produced magnetically: thus a time-varying electric field may be created by a time-varying magnetic field in a process called **electromagnetic induction**. The particular interest in this type of electric field in the present context lies in the fact that the induced field forms closed loops. Such an electric field will induce circulating currents in a conducting fluid by driving mobile charges along the field direction. Now, currents that flow in closed loops cannot lead to charge accumulation; hence this type of applied electric field is not cancelled to zero within a conducting fluid.

So overhead high-voltage power lines may exert some effect through the alternating magnetic field they create because the alternating fields generate current loops within the specimen. Electric field shielding by the conducting fluid would still operate beneath cells, as is illustrated in the following example. Consider a fluid-filled cylindrical culture dish placed within the field of a large cylindrical solenoid so that the axes of the dish and the coil are coincident. Let the solenoid generate a magnetic field that has a uniform intensity over the dish but which oscillates in time. Electric fields are then induced in the fluid contained in the dish in the form of circles around the common axis as shown in Figure 6.9. For cells within the fluid, the induced electric fields in the fluid are external fields and so will be excluded from the interior of the cells by an accumulation of charges on the inside surfaces of the cell membranes. The induced currents will therefore flow round the cells rather than induce currents within them.

Thus, although the electric fields produced by externally applied alternating *magnetic* fields do penetrate conducting bodies more effectively than externally generated *electric* fields, they are still very effectively excluded from the cell interior. Moreover, the situation today is that, despite much experimental activity and many speculative suggestions for possible mechanisms, there are no experimentally confirmed examples of the interaction of small-amplitude externally applied electric fields and living organisms at the cell level.

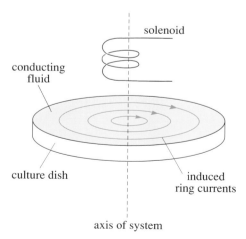

**Figure 6.9** Circulating currents induced in a conducting fluid.

## Objectives for Chapter 6

When you have completed this chapter, you should be able to:

6.1    Define and use, or recognize definitions and applications of, each of the terms printed in **bold** in the text.

6.2    Outline the significance of symmetry breaking when the previously identical cells in a developing embryo start to differentiate, and describe one of the mechanisms of symmetry breaking.

6.3    Describe the principles and uses of one non-invasive experimental technique for each of the following:

(a) the detection of internal magnetic fields and hence electrical activity produced by living organisms.

(b) the use of external electromagnetic fields to elucidate internal structure in living organisms.

6.4    Explain, with examples, the importance to some animals of being able to detect the direction of the Earth's magnetic field, and show how the presence of magnetite crystals within the structure of an animal may make such detection possible.

6.5    Describe qualitatively how the conducting fluids within living animals tend to exclude externally generated electric fields.

## Questions for Chapter 6

### Question 6.1 *(Objective 6.3)*

An MRI scan is taken of a subject's head. The proton (hydrogen atom) distribution in slices that are 1 mm thick in the $x$-direction is to be determined. At the radio-frequency at which the MRI machine operates, the proton resonance is detected within a range of magnetic fields which differ from the resonance field by $\pm 10^{-4}$ T. Estimate the minimum amplitude of the magnetic field gradient that must be applied along the $x$-direction.

### Question 6.2 *(Objective 6.5)*

Comment briefly on any possible significance for living systems of the fact that, while static magnetic and electric fields occur naturally, radio-frequency oscillating, electric and magnetic fields are almost exclusively the products of human technology.

## CHAPTER 7

# Biological rhythms and biocommunication

## 7.1 Introduction

The recognition that organisms exert very precise control over their internal states, as indicated by the near-constancy of crucial quantities such as body temperature and the oxygen content of the blood, gave rise to the basic concept of homeostasis. If there is too much variation in such vital physiological variables, the organism comes to grief – for example, the brain can be damaged from oxygen deficiency or the organism may die from a fever in which the body temperature rises above a critical value (around 41 °C or 106 °F in humans). However, a more detailed study of physiological state reveals that such quantities are not constants after all; they have well-defined patterns of variation in time. Your body temperature, for instance, is not constant at 37 °C but varies rhythmically from a maximum in the evening to a minimum in the early morning with a difference of about 0.7 °C. Daily rhythms of this kind occur in most physiological variables that were previously considered to be regulated at fixed values, and a host of other rhythms with different periods, ranging from fractions of a second to many years, have been identified in organisms and populations. Evidently organisms are not homeostatic, but homeodynamic systems, as emphasized in Chapter 1 of Book 1. The task of this chapter is to take a look at the evidence for this dynamic characteristic of life, to develop an appreciation of what a remarkable range of rhythms occurs in organisms, and to understand both the causes and the significance of this rich pattern of temporal organization in living systems. As usual in biology, this requires that we examine both the dynamic origins of the rhythms, that is, what causes them; and also that we try to understand in what way rhythmic behaviour serves organisms – what opportunities it presents for developing integrated states of organization within the organism and with the environment. The relations between rhythmic activity and biocommunication at all levels of biological organization will then become evident.

### 7.1.1 Organisms are rhythmic systems

If you have been sitting quietly for a few minutes, take your pulse and record it in the usual form of beats per minute. Now do the same with your respiration rate and record the number of full inhalation–exhalation cycles per minute. Divide pulse rate by respiration rate.

▷ What ratio do you get?

▶ The usual average for this is about 4 : 1.

Notice that it is easier to be confident about your pulse rate than your respiratory rate.

▷ Why is this?

▶ Because you exert voluntary control over your respiration but not over your heart beat, unless you are trained in yoga or have developed relaxation and stress reduction techniques. So to get a good measure of respiration rate you need to repeat it a number of times and take the average, which is of course what should be done for any reliable measurement.

Pulse and respiration rate are correlated with one another in a flexible manner, but tend to lock into integral ratios such as $3:1$, $4:1$, or $5:1$ when your physiology has settled to constant conditions of sleeping or sitting or walking. Similar correlations arise between other rhythms. You will notice that your breathing tends to fall into harmony with the rhythm of walking or running so that you shift through discrete breathing frequencies and establish particular ratios of one rhythm to another as you change pace and feel the demand for more oxygen. Athletes try to get the ratio of breathing to running pace down by developing greater lung capacity. The whole rhythmic pattern of physiological activity has the character of a multi-mode system that can shift over a spectrum of frequencies in a variety of rhythms that tend to fall into stable, integral ratios, one rhythm comfortably nested within another.

▷ How many heart beats would you have per day if your heart continued at a steady rate of 70 beats per minute?

▶ $70 \times 60 \times 24 = 100\,800$ or about $10^5$ beats per day.

This ratio of around $10^5$ between the frequencies of the heart and diurnal rhythms does not result in any frequency correlation; the heart runs freely at its own rate without locking onto the sleep–wake rhythm. There is, however, some change in frequency of heart rate with the activity cycle and with body temperature, higher frequencies occurring during the day when you are active.

What about the other end of the biological frequency scale, the longer periods? One of the most familiar is the 28-day menstrual cycle in women, punctuated by ovulation and menstruation about 14 days apart. This fundamental physiological cycle, on which all of our lives depend in the most direct and literal sense, has been a primary focus of research for many years. As a result, it has become clear that many different levels of organization and timescales in physiological processes are coordinated within a single, integrated process. It is worth examining this cycle in some detail because a number of properties that are basic to communication, control, and rhythmic activity emerge from this example.

### 7.1.2   Regulation of ovulation and menstruation

The pituitary gland, whose activities in physiological regulation were described in Section 4.2.1, is directly involved in the ovarian cycle. It releases gonadotropic hormones into the blood, initiating the growth of a follicle in the ovary within which an oocyte matures. The growing follicle produces oestradiol, a hormone that stimulates the pituitary to produce more gonadotropin. This is a positive feedback control loop in which plasma oestradiol levels increase to a threshold that initiates a surge of gonadotropin production by the pituitary after about 14 days, triggering release of the now mature oocyte from the follicle into the Fallopian tube. The empty follicle in the ovary, now called a corpus luteum, starts to produce another hormone, progesterone, which inhibits production of gonadotropin by the pituitary. What was a positive feedback loop becomes negative feedback (see Figure 7.1). If the oocyte fails to be fertilized and no implantation of the egg into the uterine endometrium occurs, the corpus luteum continues its activity for 14 days, after which it degenerates. The pituitary then increases its gonadotropin production, initiating a new cycle. Fertilization and implantation, however, result in chemical signals from the implant that maintain the corpus luteum and suspend another cycle until the end of pregnancy.

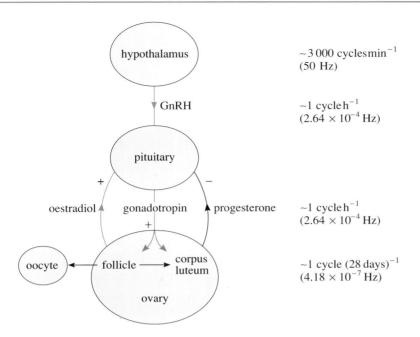

~3 000 cycles min$^{-1}$
(50 Hz)

~1 cycle h$^{-1}$
($2.64 \times 10^{-4}$ Hz)

~1 cycle h$^{-1}$
($2.64 \times 10^{-4}$ Hz)

~1 cycle (28 days)$^{-1}$
($4.18 \times 10^{-7}$ Hz)

**Figure 7.1**   Control signals linking the activities of the hypothalamus, pituitary and ovary, contributing to the regulation of the menstrual cycle.

This basic positive and negative feedback cycle between pituitary and ovary is sufficient, in principle, to produce a rhythm. The duration of the rhythm depends upon the time constants of the constituent processes – in this case, the growth and maturation time of the follicle and its oocyte, followed by the degeneration time of the corpus luteum.

The overall period of the menstrual cycle is dependent upon these processes of cell growth, differentiation, and degeneration, whose typical durations are many days. This does not explain why the period is on average 28 days, the period of the lunar month. This type of synchronization between an internal and an external rhythm raises the question whether the internal cycle is self-generating and self-maintaining, or whether it is dependent upon a periodic external stimulus. For the case of the menstrual cycle, this can be easily answered. There are many individuals whose cycles have periods greater or less than 28 days, so they must be running on their own frequencies. And in our culture moonlight is no longer a significant signal, at least in physiological terms.

▷ Rhesus monkeys also have a menstrual cycle with an average period of 28 days. How would you design an experiment to see if the lunar light signal is a synchronizing stimulus?

▶ You could enclose monkeys so that they don't see the Moon, and compare their rhythms with a control group that is not enclosed. The former should lose synchrony with the Moon and run at other frequencies.

▷ What conclusion would you draw if a group of females enclosed together all become synchronized to a cycle of 27 days, menstruating at more or less the same time?

▶ Possibly the group synchronize to one another, locking onto a common frequency.

Experiments of this type have shown that groups of monkeys not exposed to the Moon do synchronize to a common rhythm other than 28 days. When separated from each other, the synchrony is lost and individuals run freely at different frequencies. The same phenomena have been observed in humans.

If the cycle of ovulation was as simple as that described above, with positive and negative feedback phases of the interaction between pituitary and ovary driving the rhythm, its details would no doubt have been worked out by now. The story is in fact much more complex. First, it emerged from close monitoring of the blood gonadotropin levels that the pituitary does not simply produce these hormones in a cycle that rises and falls with a 28-day period. The endocrine signal from pituitary to ovary is periodic, but it has a frequency of about one pulse per hour. Furthermore, by modulating this signal (in rhesus monkeys) it was shown that if this period is 90 min instead of 60 min the follicle does not respond and the cycle fails; and similarly there is failure if the gonadotropin cycle is too short (<30 min). So there is a frequency window, a narrow range of gonadotropin pulse frequencies, to which the follicle can respond by growing and maturing.

What is the origin of the pituitary rhythm? It has been known for some time that gonadotropin release from the pituitary is dependent upon another endocrine signal, a gonadotropin-releasing hormone (GnRH) that is produced by the hypothalamus, deep in the brain stem. GnRH is produced and released rhythmically about once every hour, and the pituitary rhythm is dependent upon this. So the source of the high-frequency signal (relative to menstrual cycle duration) that controls ovulation is, so far as is currently known, the hypothalamus (Figure 7.2).

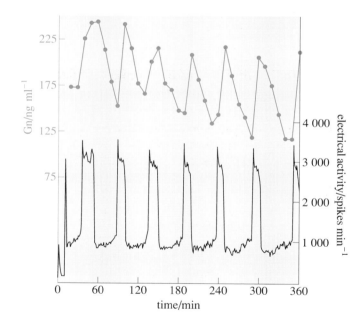

**Figure 7.2**  Rhythmic electrical activity of the hypothalamus (lower trace, electrical spikes per minute), correlating with gonadotropin concentration in the blood (upper trace), mediated by GnRH (not measured).

There is an intimate connection between the electrical activity of the brain and its endocrine functions. So it might be expected that an electrical signal accompanies the pulsatile release of GnRH from the hypothalamus, and indeed this has been observed. Figure 7.2 shows a direct correlation between spikes per minute recorded from the thalamus and plasma gonadotropin levels in a rhesus monkey, the electrical activity varying rhythmically from about 1 000 to 3 000 spikes per minute. It is inferred from this that the periodicity of the electrical signal is causally connected with pulsatile GnRH release from the hypothalamus, though this is not shown directly. What we see here is something characteristic of electrical communication: the information in the signal which initiates the response is carried by the frequency, involving frequency modulation.

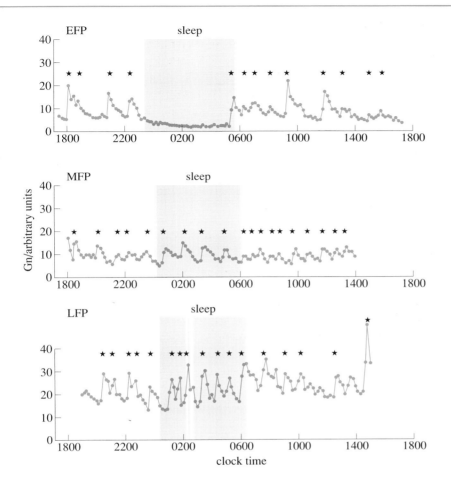

**Figure 7.3** Pulsatile gonadotropin concentration (measured in arbitrary units) in the blood of volunteer subjects during the course of a day in the early (EFP, top), mid (MFP, middle) and late (LFP, bottom) follicular phase of the normal menstrual cycle.

▷ Assuming that the 1 cycle per hour pulse of gonadotropin continues at the same frequency throughout the duration of follicle maturation (14 days), how many pulses will the follicle receive?

▶ $24 \times 14 = 336$ pulses

▷ What major biological rhythm lies between 1 cycle h$^{-1}$ and 1 cycle per 28 days?

▶ The sleep–wake rhythm of 1 cycle day$^{-1}$.

As one might expect, the sleep–wake cycle modulates the endocrine rhythm, but its influence varies with the time of the menstrual cycle. This is shown in Figure 7.3, which shows data from 36 women menstruating normally who volunteered to have blood samples taken every 10 minutes over a period of 24 hours. In the early follicular phase of the cycle (EFP), the pulses of gonadotropin virtually cease during sleep. In middle and late phases of this 14-day follicular maturation period (MFP and LFP respectively), the pulses continue during sleep, and build up in amplitude during the late phase. Notice that the pulses are actually quite irregular. It will become evident later that this is characteristic of physiological rhythms. Even your heart beat, apparently so steady, is actually quite irregular, but irregular in a characteristic way that reveals the occurrence of *chaotic dynamics* as a component of the periodic process. Paradoxically, very regular rhythms are diagnostic of potential trouble, as will emerge in Section 7.3. Chaos turns out to be a source of stability!

▷ From Figure 7.3, what is the average frequency of the pulses in the MFP, taking a star to identify a pulse?

▶ 1 pulse every 1.053 h (19 pulses in 20 h).

▷ Convert this to pulses per second (hertz, or Hz), which is the standard frequency measure.

▶ $1/1.053 \times 60 \times 60 = 2.64 \times 10^{-4}$ Hz.

During the luteal phase of the cycle, when the corpus luteum produces progesterone and inhibits pituitary gonadotropin production, the frequency of gonadotropin release decreases but does not entirely cease, even during the late stage of this phase. However, as noted above, the narrow frequency window of follicle response to gonadotropin means that a frequency of less than 1 pulse every 90 minutes fails to stimulate follicular growth and the pituitary signal frequency falls below this value during the luteal phase. Evidently the signal from the pituitary to the ovary also functions by frequency modulation of a rhythmic control signal, with response tuned to a narrow frequency range. This is not the conventional picture of biological control systems, in which steady concentrations of chemicals are regarded as the signals that determine different responses, depending on whether they are above or below a critical threshold level. However, this classical view can be combined with frequency modulation by adding the other variable into the calculation, namely amplitude, as seen towards the end of the follicular phase when pulse amplitude is significantly increased, leading up to the surge that triggers ovulation.

These observations show how the original concept of homeostatic control turns into homeodynamic regulation, both frequency and amplitude modulation emerging as significant components of biological signal transmission and reception systems. Any electronic engineer will tell you that these are the variables to use in designing reliable, robust control and communication systems. The menstrual cycle turns out to be a highly sophisticated control system with component frequencies varying from 3 000 oscillations min$^{-1}$ (50 Hz) in the electrical activity of the hypothalamus to 1 cycle per 28 days ($4.13 \times 10^{-7}$ Hz) in the ovary. This is eight orders of magnitude in the frequency range, coordinated into a single physiological process. The frequencies are nested within one another, resulting in a system with hierarchical organization in which processes occurring on different timescales are coordinated through frequency and amplitude changes in higher frequency signals, which cause rhythmic modulations of state in organs with slower response times. This type of dynamic organization results in both stability of rhythmic operation and flexibility in the frequencies of the constituent processes.

Hierarchical dynamic organization of this type is a good design for evolving systems because it produces reliable performance while allowing for modifications at each level of activity, including the addition of new levels of activity at either end of the hierarchy. It is also a natural, robust pattern of order to emerge in a complex dynamic system. Other frequencies within the range encompassed by the components of the menstrual cycle will be encountered as different systems are examined in this chapter. However, just to get an idea of the full range of the rhythmic spectrum that organisms make use of, it is of interest to mention two other examples.

The first of these are biological cycles with periods of about one year. Seasonal variations are very widespread, especially in latitudes where temperature and day length vary significantly during the course of the Earth's annual journey around the Sun. Animals shed their winter coats and then grow them again; birds migrate; trees shed their leaves and grow them again. Human physiology also reveals annual rhythms, despite our relative insulation from seasonal change. For instance, the content of certain steroids in the urine (17-ketosteroids) in a group of North American men varied rhythmically from a peak in November to a trough in May. Many other physiological variables show such annual rhythms. Also, studies carried out in France on the incidence of death from such causes as cerebral vascular lesions, which would appear to have no obvious relationship to season, revealed a rhythm with a peak in January and a trough in August. Again the question arises whether such rhythms are endogenous or exogenous, or a result of synchronization between internal and external periods. This is difficult to answer in humans, but studies on cave-dwelling crayfish in the Northern Hemisphere suggest that the rhythm of breeding condition in males, which peaked in December and fell to a minimum in July, is endogenous. So it appears that some combination of interacting signals can generate biological rhythms with a frequency of about 1 cycle per year ($3.17 \times 10^{-8}$ Hz). Clearly this must involve processes with long time constants such as growth and maturation of tissues of the type that occurs in the ovarian cycle but some ten times slower. This is certainly not improbable, especially when the next example is considered.

The prize for extended duration of a cycle must surely go to those species of bamboo whose life cycle has a duration of about 120 years! This is the time for the bamboo plants to grow from seed to a condition of flowering, which they do only once, producing seed that germinates to produce the next generation. If indeed the biological race for survival goes to the swift, then this appears to be a precarious strategy. However, it works perfectly well. All that is required is that one generation replace the next, and so long as there aren't too many pandas eating them before the new plants get established, it is an adequate strategy. The panda also has a precarious life-style. Its digestive system is unable to make use of the cellulose which forms the main bulk of its diet, so it has to spend much of its life eating bamboo shoots to satisfy its nutritional requirements. After the bamboos flower and produce seed, the plants die back and there are no shoots available until the new generation of plants has germinated. Since the bamboos in a local region tend to be synchronized in their flowering, the pandas have to roam a sufficiently large territory not to be caught out by a sudden interruption in their favourite diet.

## Summary of Section 7.1

Physiological regulation is achieved by a dynamic process that frequently involves rhythmic activities rather than constancy of physiological variables, as implied by the concept of homeostasis. Rhythms with different frequencies can interact with one another to produce stable periodic patterns, many of which match up with environmental periodicities (diurnal, lunar, annual, etc.). The range of biological frequencies is large, from 17–50 Hz for electrical activity in the brain (see Figure 7.2) to 1 cycle per 120 years for the life cycle of some bamboo species.

## 7.2 The dynamics of rhythms

### 7.2.1 What makes clocks tick?

Two natural rhythms dominate our sense of time: the rotation of the Earth around the Sun and of the Earth on its axis, giving rise to two of the frequencies already encountered, the annual and the diurnal. On this basis we have constructed a sense of time and its measurement, replacing the natural cycles by our own technology and sub-dividing the natural periods into sub-units of hours, minutes, seconds, and so on down to picoseconds ($10^{-12}$ s) and beyond. For really accurate time-keeping we use atomic clocks, another source of natural rhythms whose frequencies are in the range of $10^{14}$ Hz. Do all rhythmic systems involve a common dynamic property that can be regarded as the cause of their periodic behaviour or are there basic differences between them, their repetitive cycling being the only common feature? In this section, the objective is to understand what type of dynamic organization gives organisms their rhythmic properties. To get at the basic principles involved, it is useful to start with the traditional pendulum clock and then to consider in what ways organisms are similar to, and in what ways they differ from, this type of dynamic system. (There is no need to remember the equations in this section.)

The swinging pendulum has been *the* paradigmatic example of rhythmic activity ever since Galileo watched the hanging lamps swinging in the churches of Pisa and realized how he could analyse their motion in terms of his new physics. But to do this he had to take a step that was far from obvious: he had to ignore that fact that a real pendulum always comes to rest if left undisturbed, and describe it as if it were in perpetual movement. So he invented the idea of a frictionless, ideal pendulum, just as he introduced the idea of objects falling through a vacuum in which a stone and a feather fall at the same rate. Such are the abstractions of mechanics.

The motion of the ideal pendulum is shown in Figure 7.4. The arm of the pendulum of length $l$ (assumed to have no mass) is displaced from the vertical position through an angle $\alpha$ and then released. The mass of the bob, $m$, moves under the action of gravity, $g$, falling along the arc of the circle. At any moment the arm of the pendulum is at an angle, $\theta$, to the vertical, varying between $+\alpha$ and $-\alpha$ at the extremes of the swing. The mathematical problem is to find an expression for the precise way in which $\theta$ varies with time so that the motion is exactly described. Even when friction is ignored, this is a difficult problem. However, there is one further simplification that makes the problem simple. If it is assumed that $\alpha$ is small, so that $\theta$ is always small, then the equations become those for the simple harmonic oscillator, and can readily be solved. The solution is:

$$\theta = \alpha \cos \sqrt{\frac{g}{l}} \times t \tag{7.1}$$

where $t$ = time, $g$ = acceleration due to gravity, $l$ = length of the pendulum and $\alpha$ = initial displacement.

Then $\theta$ increases and decreases smoothly with a regular rhythm, called **simple harmonic motion**, described by the cosine function with frequency

$$\sqrt{\frac{g}{l}}$$

The *period* of the motion is

$$2\pi \times \sqrt{\frac{l}{g}}$$

This is a very useful expression. It tells us that the period can be altered by changing $l$, the distance from the fixed pivot to the mass. This is how the period

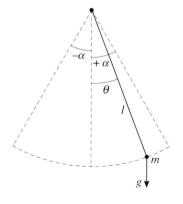

**Figure 7.4** The periodic motion, starting from a displacement angle $\alpha$, of an ideal pendulum of length $l$ and mass $m$, acted upon by gravity ($g$), with motion measured by the angle $\theta$ between the pendulum and the vertical position.

of a real pendulum in a clock is adjusted, by altering the position of the weight on the pendulum arm. The relationship between period and length, $l$, is not exactly a square root function for an actual clock, but it's pretty close – especially for a grandfather clock with a long pendulum and a small angle of motion, $\alpha$. The expression can also be used to estimate the force of gravity, $g$.

▷ Will a pendulum of fixed length, $l$, have a larger or a shorter period on the Moon as compared with the Earth?

▶ $g$ is smaller on the Moon, so the period will be longer. The expression actually allows for an estimate of $g$ for a pendulum of known length $l$, simply by measuring the period of oscillation.

Notice that there is no contribution of the mass, $m$, to the period. This is a result of the assumption that there is no friction, a perfect bearing for the arm, and no air resistance. The same assumptions result in a stone and a feather falling at the same rate through a vacuum.

Equation 7.1 is clearly very useful, but there is another way of looking at the motion of a pendulum that is even more informative. This is based on its energy. Newton's laws of motion lead to the law of conservation of energy, which states that the sum of the kinetic and the potential energy of a moving body is a constant. This is a result of the assumption that there is no friction, which would gradually turn energy into heat, which is dissipated (see the discussion of the laws of thermodynamics in Book 2, Chapter 2). Calling the velocity $v$ and distance of the pendulum from the rest position $x$, the energy can be expressed as:

$$E = \frac{m}{2}\left(v^2 + \frac{g}{l} \times x^2\right) = \text{constant} \tag{7.2}$$

where the first term is the kinetic energy, and the second term is the potential energy. As the pendulum swings, potential energy is converted into kinetic energy and back again. Dynamic systems of this type are called energy-conserving or **conservative systems**.

The expression for the energy doesn't involve time, but it can be used to describe the motion of the pendulum geometrically. To do this, it is convenient first to change the variable $x$ into $u$ by writing

$$u = \sqrt{\frac{g}{l}} \times x$$

Then the energy takes the form:

$$E = \frac{m}{2}(v^2 + u^2) = \text{constant}$$

$E$ is now a sum of squares of $u$ and $v$, so if we plot the curve describing $E$ in terms of $u$ and $v$ we get a circle. The radius of the circle, which describes the *amplitude* of the oscillation, expresses the amount of energy in the pendulum, which depends upon the angle of swing, $\alpha$, as well as upon the mass, $m$. Different energies are then described by circles of different radius, as in Figure 7.5.

The motion of the pendulum can be represented by a point moving around one of the circles in Figure 7.5. We can think of the pendulum starting from a position of maximum displacement to the right, where $\theta = \alpha$ in Figure 7.4, and $u$, the distance to the vertical position, is a maximum while $v$, the velocity, is zero. This is the point where a circle meets the $u$ axis at a positive value, say point 1. As the pendulum swings, $u$ decreases and the velocity, $v$, increases, the point describing the motion moving along the arc of the circle in the direction of the arrow. At point 2 the pendulum has reached the vertical position ($u = 0$) and the velocity, $v$, is a maximum. Then the pendulum swings to the left of the vertical so that $u$ becomes negative while $v$ decreases, reaching zero at the top of the swing on the left where the pendulum changes its direction and $u$ takes its maximum negative value, at point 3. The motion then reverses and the curve passes through point 4 on its way back to 1. So each cycle of the pendulum is now described by one complete circuit of the circle, whose constancy describes geometrically the law of conservation of energy for the frictionless motion. Each circle is called a trajectory, describing the motion. The position of a point on a circle, measured by the angle, $\phi$, from the $u$-axis to the radius connecting the origin to the point, is called the **phase of the motion**.

▷ A pendulum is resting vertically with no motion. How is this described in Figure 7.5, and what is its energy?

▶ Since for this pendulum $u = v = 0$, it is described by the point at the centre of coordinates and the energy is zero: the circle has zero radius.

▷ A real pendulum with friction loses energy. Suppose it starts at position 1 in Figure 7.5. Eventually this pendulum will come to rest at $u = v = 0$, with no energy. What curve describes its motion between point 1 and the origin?

▶ The curve that describes the motion of a real pendulum is a spiral, like that shown in Figure 7.6a. The more the friction, the more rapidly will the spiral converge on the origin. This stable point of the motion is called a **point attractor**, for obvious descriptive reasons, and the spiral path is the **trajectory** for the pendulum starting at point 1 and ending at point 0.

To get a real clock, it is necessary to alter the spiral trajectory of Figure 7.6a so that there is a closed cycle. The usual way of doing this is to add a bit of energy to each cycle, which is described by the short, broad vector marked $e$ in the figure. This is the role of the escapement mechanism, which gets its energy from the spring. The result is that the spiral turns into a closed cycle, as shown in Figure 7.6b. You might wonder how a clock-maker designs an escapement mechanism to add exactly the right amount of energy to balance what is lost in one cycle. However, this is where natural processes make things easy. The escapement mechanism needn't be precise in the energy it adds, for the whole system will settle to a particular cycle for any small energy addition to each circuit. This is

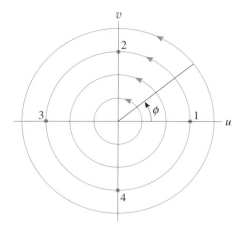

**Figure 7.5**   The motion of an ideal pendulum in terms of variables $u$ and $v$, which describe the motion as a closed circle whose radius squared is proportional to the energy (see text for details).

**Figure 7.6**   (a) The motion of a real pendulum with friction: the energy gradually decreases until the pendulum comes to rest at the origin. (b) The motion of a real clock pendulum with an escapement mechanism that adds an impulse of energy once per cycle, producing stable motion.

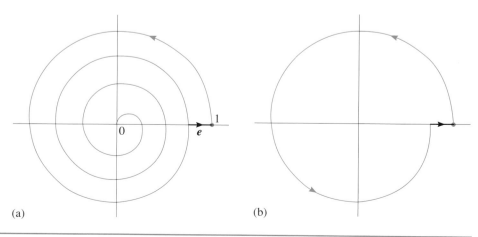

(a)                              (b)

because the amplitude of the oscillation varies with the amount of energy added, and so does the amount of energy dissipated in each cycle. So for any added increment that is small the pendulum will always settle to some cycle. The clockmaker then has the problem of getting the clock to keep accurate time. This is most easily done by altering the length of the arm, $l$, as described earlier.

Once all the necessary adjustments have been made, the clock settles down into a stable cycle of the type shown in Figure 7.6b. If there is a disturbance to this cycle, such as someone accidentally hitting the clock slightly, then there will be a brief disturbance of the motion but it will settle down again to the same periodic cycle, so long as all the parameters remain constant: $m$, $l$, $g$, and the amount of energy, added to each cycle by the escapement mechanism. Therefore the closed cycle of Figure 7.6b is now an attractor for the dynamics of the clock – that is to say, any neighbouring trajectory ends up on this cycle.

In the literature on dynamical systems, such a periodic attractor is known as a **limit cycle** (it is the cyclic attractor to which a whole family of trajectories converges in the limit of increasing time). These occur only in open systems that consume and dissipate energy (recall the definition of open as compared with closed systems in Book 2, Chapter 2). Because they don't conserve energy, these are called **non-conservative oscillators**. Biological clocks do not use escapement mechanisms to get stable periodic attractors, but they do dissipate energy, needing a continuous energy input to keep them going. So they are open and non-conservative.

## Coupled oscillators

There is a story that the famous 17th-century Dutch physicist and mathematician, Christiaan Huygens, who made very important contributions to the study of light, was in a clock shop in Leiden one day when something suddenly attracted his attention. A number of clocks were on display on a wall and he noticed that they were ticking with the same frequency. He waited and listened, expecting them to fall out of synchrony after what he assumed was a momentary coincidence.

However, they continued with the same beat. They were all clocks of about the same size with pendula of similar lengths, so it was to be expected that they would have approximately the same frequency. But for them to be synchronized to exactly the same rhythm was remarkable. Furthermore, similar clocks mounted on other walls were not synchronized. Huygens then noticed that the synchronous clocks were mounted, not on a solid wall, but on a fairly thin board. He deduced that what must be happening is that the mechanical vibrations of the ticking clocks were transmitted to one another through the thin board, which was itself vibrating.

This was one of the earliest reports of synchronization of oscillators by mutual coupling, which is a particular form of **resonance** between rhythmic processes. Just as the escapement mechanism of a real clock adds a little bit of energy and produces a stable periodic cycle, so two or more clocks that are exchanging small amounts of energy with each other (resonating) can settle into synchrony taking the same rhythm though not necessarily the same phase – i.e., they don't, in general, tick together, but they maintain a constant phase difference so that the interval between ticks is constant. This type of synchronization is also called mutual entrainment of oscillators to constant frequency. It occurs only with non-conservative oscillators.

The phenomenon of the follicle in the ovary responding only to a narrow frequency range of gonadotropin pulses, described in Section 7.1.2, is another example of resonance. In this case the resonance is not between two coupled oscillators but between a periodic signal (the gonadotropin pulses) and a target organ that is not itself oscillating independently. This type of resonance is actually very familiar. If you have ever had a poorly aligned wheel on your car that produces a small vibration, then, within a narrow range of speeds, when the wheel is vibrating in a certain frequency range, you will have found that both the steering mechanism and the whole car start to vibrate with a rather alarming amplitude. This is because the steering mechanism has its own characteristic response time to a disturbance, and if a periodic vibration is within the right frequency range the mechanism resonates with this response time, producing a large-amplitude periodic vibration. This is known as a forced oscillation, the system responding at a characteristic frequency which in general is not the same as the forcing frequency, and is usually smaller by some whole number ratio (a rational fraction) such as 1/4 or 2/5. This is called subharmonic resonance. Exactly what type of resonance is occurring in the follicle to the pulses of gonadotropin from the pituitary is not yet known.

An important instance of synchronization is when there is a separate external stimulus that acts periodically on an oscillator or a set of oscillators, which need not be coupled to one another. Such a stimulus can entrain the oscillators to its frequency. An example of this was described in Section 7.1.2 in which rhesus monkeys had their menstrual cycles synchronized to the Moon, which acted as the entraining stimulus. The monkeys can also become synchronized simply by behavioural interaction, which is entrainment by mutual coupling as in Huygens' clocks.

### Summary of Section 7.2.1

The motion of a real pendulum is not easy to describe in precise mathematical terms, but a good approximation to its motion, called simple harmonic motion, is extremely useful and involves the concept of energy conservation. A real clock does not conserve energy, requiring periodic impulses from an escapement mechanism to produce a stable periodic trajectory, a cyclic or periodic attractor. Biological oscillators are also non-conservative, requiring metabolic energy to drive them, in common with all living processes. These oscillations can interact with other periodic processes, resulting in a state of resonance or entrainment in which the different oscillators have a stable relationship. This is the basis of the entrainment of biological oscillators to environmental frequencies, and the emergence of synchronous behaviour in interacting organisms.

### 7.2.2  Circadian organization: cellular rhythms

Rhythmic activity with a periodicity of 24 h is a virtually universal property of organisms, from unicellular organisms to whales. One of the first species to have its daily cycles systematically investigated in the laboratory is a little unicellular marine dinoflagellate with the imposing name of *Gonyaulax polyedra*. This is an alga that lives in the Caribbean Sea and one of its distinguishing characteristics is its luminescent activity, which is particularly marked when it is mechanically disturbed, as by wave action.

Studies by Beatrice Sweeney of the Scripps Marine Laboratory in California and J. Woodland Hastings at Harvard University on laboratory cultures of these cells showed that in these cultures there is in fact a daily rhythm of spontaneous, flashing luminescence. There is also a rhythm of steady glow, the intensity rising to a peak early in the morning just before dawn and then subsiding during the day.

An initial question concerned the source of this rhythm: was it due to the daily light–dark cycle, or did it have an endogenous origin? This was examined by putting the cultures under constant conditions of temperature, light and all other controllable environmental variables. Since these algae are obligate photoautotrophs, having to make their own nutrients by photosynthesis and therefore requiring light to survive, these constant conditions had to be continuous light. With dim light, the rhythm of luminescence continued indefinitely, revealing an endogenous oscillation. Moreover, the period of this oscillation was not 24 h, but about 24.5 h at 21 °C, which is further evidence for an independent oscillator in the cells. Such oscillations with periods near 24 h are called **circadian rhythms**, from *circa*, about, and *diem*, day. The daily light–dark cycle entrains this endogenous rhythm to 24 h.

When you think about it, this is a very sensible way to get a biological clock to run on a precise 24 h cycle. Designing one in an organism that runs with exactly this period would be a formidable task, and quite unnecessary if there is a convenient synchronizing signal to hand, such as dawn and/or dusk. The difficulty of designing a precise time-keeper is evident from the fact that the British Admiralty had to offer £20 000 (in 1714 currency!) as the reward for a sufficiently precise ship's chronometer to allow accurate navigation. It was 45 years before an adequate design was submitted. John Harrison's fourth attempt produced a chronometer that had an error of only 1/17 000 of a cycle, corresponding to a navigational error of $1\frac{1}{2}$ miles at equatorial latitudes, on a test during a stormy voyage to the West Indies in 1759. Organisms use an easier method, with an endogenous oscillator whose period is not far from 24 h that is entrained by the diurnal light–dark cycle (called a *Zeitgeber* or time-giver). Why didn't John Harrison think of using an organism such as *Gonyaulax* for a ship's chronometer? Storms would not disturb its rhythms in the least. But science wasn't tuned in to such aspects of biology in the 18th century, and mechanical devices were regarded as the only reliable measuring instruments.

How does an organism produce a rhythm with a period of near 24 h in the first place? As you saw in Section 7.1.1, biological control circuits are prone to oscillate because of positive and negative feedback control loops. So organisms have a tendency to produce oscillations and these will have a great range of frequencies because of different time constants in the constituent processes. A number of these have already been mentioned in this chapter. Electrical activities are fast, as in the 50 Hz oscillations of the hypothalamus (Figure 7.1); metabolic processes are slower, resulting in rhythms with periods in the range of a few minutes to hours, as in the pituitary pulses; growth and differentiation of tissues takes days, and you have seen that these are involved in the 28-day menstrual cycle. In between are circadian rhythms. These are the oscillations that happen to fall close to the frequency of 1 per 24 h, though to qualify as reliable chronometers these oscillations must have some distinctive properties, as will emerge in this section and in Section 7.2.3.

Observations on the *Gonyaulax* luminescence rhythm have revealed interesting properties of these clocks, important for understanding the dynamics of biological oscillators and biocommunication. The results of a typical experiment with a culture of cells is shown in Figure 7.7. The culture was initially grown under conditions of 12 h of light followed by 12 h of darkness (LD 12:12), then transferred to constant dim light (DD) for the duration of the record presented. Time in hours is plotted along the abscissa (horizontal axis), while the intensity of the glow rhythm in arbitrary units is plotted up the ordinate (vertical axis). The temperature throughout was 19 °C.

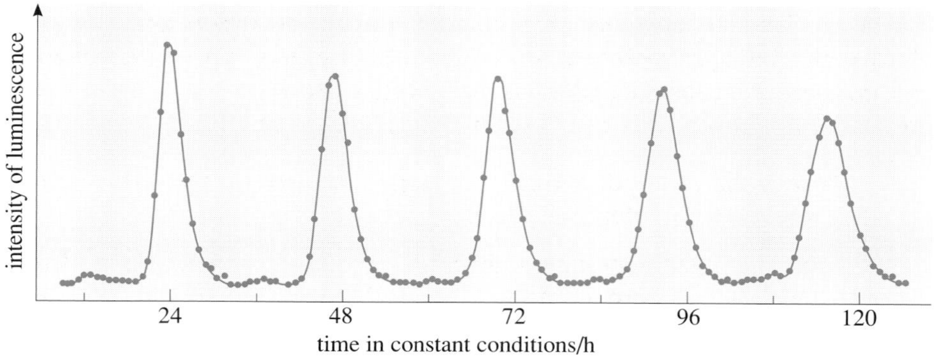

**Figure 7.7** The intensity of the luminescent glow from a culture of *Gonyaulax* cells as a function of time, showing a circadian rhythm.

▷ The amplitude of the peaks decreases and the duration of the glow phase increases during the first few cycles shown. What does this mean about the synchrony of the culture?

▶ The degree of synchrony evidently decreases when the culture is transferred from LD to DD.

▷ If the cells became completely desynchronized in DD, so that each individual cell continued to have the same circadian period but the cells lost their phase coherence, what would the glow pattern of the culture look like?

▶ It would be flat, at some value above the troughs in Figure 7.7.

The culture does lose some of its initial synchrony on transfer from LD to DD, but it levels off at a fixed value on the ordinate. *Gonyaulax* are free-swimming, flagellated cells that are not in intimate contact with one another. They must be signalling to one another, either by their own light signals or by releasing something into the seawater at particular phases of the cycle. Clearly the cells have considerable sensitivity to one another's phase to remain mutually entrained to a population rhythm.

▷ What is the mean period of the cycles?

▶ 23 h.

▷ At 19 °C the cells are cycling faster than at 21 °C, where the period was 24.5 h. Is this what you would expect?

▶ Generally speaking, chemical reaction rates decrease as temperature decreases. If biological clocks depend upon chemical reactions, then one would expect them to run more slowly at lower temperatures, but in this instance the opposite is evidently the case.

Most circadian rhythms do run more slowly at lower temperatures, but the effect of temperature on rate is much less than expected. This is called **temperature compensation**, and it keeps the clock period close to 24 h, independently of ambient temperature. This is a property required of a good chronometer, maintaining the oscillation well within the entrainment limits of the 24 h day/night signal.

The *Gonyaulax* cells grow and divide. As you might expect, cell divisions also have a 24 h rhythm, dividing cells being seen in the culture only towards the end of the luminescent period, at dawn. But the doubling time of a culture, which is determined by the mean generation time of its cells, is not 24 h but, at 21 °C, it is 36 h. Evidently there is a limited phase range of the daily cycle when cell divisions can occur. Even if a cell is ready to divide, it has to wait for the opening of the pre-dawn pendulum 'gate' before it can divide, and it is the circadian clock that opens the gate, and then closes it again. The result is that division is not itself rhythmic in these cells; its timing is regulated by the circadian clock.

▷ If the doubling time of a culture is 36 h, what is the expected pattern of division of an individual cell and its progeny?

▶ Suppose the experiment starts with the observation of a cell in a culture on day 0 that has just been produced by division. This cell is observed to divide on day 1, 24 h later, producing 2 progeny. On day 2 neither of these cells divides, but both divide on day 3, 48 h after the last division and 72 h after the observation period started. The average interval between divisions is then (24 + 48)/2 = 36 h; that is, there have been 2 generations in 3 days, 4 cells being produced in 72 h from the originally observed cell. There are also all the permutations of this: the cell that has just divided could wait 48 h before dividing again, with the next interval to division being 24 h; or the two progeny cells could have different patterns of 24 and 48 h intervals between divisions, and so on. Whatever the pattern, we need simply to give an *average* generation time of 36 h for the culture as a whole.

Cell division *rates* vary in the expected way with temperature, cells growing and dividing more rapidly at higher temperatures, but the *timing* of division is circadian. So it is evident that we are dealing here with rather different types of biochemical process, though there is interaction between them since the clock controls cell division. In fact, much of cell physiology is regulated by the circadian oscillator. The photosynthetic machinery gets into gear during the pre-dawn phase and more or less shuts down at night. The cells also move up and down from deeper levels to the ocean surface and back again with a daily cycle. And of course light is not the only diurnal variable; there is also temperature. However, the light–dark cycle is the most regular and reliable, not only for the 24 h cycle but also for seasonal cycles, since day-length varies in a systematic, regular way throughout the year in temperate latitudes. It is these changes in photoperiod that provide migrating bird species with the accurate information about the time of year to initiate their migratory flights. It is the subtle but regular influence of changing day-length on the amplitude and the phase of circadian cycles that acts as the signal for physiological change, the circadian rhythm continuing to play a central role in behaviour. So let us see if we can go a little deeper into the origins and the dynamic characteristics of this basic oscillator.

*Summary of Section 7.2.2*

The 24 h clock in organisms is an oscillation with a period close to 24 h that gets entrained to the Earth's day–night light cycle. The oscillation is temperature-compensated so that variations in ambient temperature have little effect on period, the oscillation remaining within the entrainment range of the diurnal cycle. Cell division and many biochemical activities are influenced by the biological clock in such a way that cellular processes are organized on a 24 h cycle.

*7.2.3   Drosophila rhythms*

The little fruit-fly, *Drosophila*, has not only made significant contributions to our understanding of genetics. It has also revealed one of the most significant properties of circadian rhythms: they can be stopped dead without any damage to or abnormality in the clock mechanism itself. What this gives is an insight into the dynamics of many other biological rhythms, among them the heart beat, to which the expression 'stop dead' has an altogether more ominous meaning. Cardiac arrest appears often to be due to similar intrinsic dynamic properties that allow the *Drosophila* circadian clock to be arrested by an innocuous but carefully timed stimulus, as will be discussed in Section 7.3.3.

The rhythm that was used to study the dynamics of the *Drosophila* circadian rhythm was the emergence of the adult fly from the pupal case. Larvae are light-sensitive and so are the pupae. The standard experimental procedure for studying emergence or **eclosion**, to use the technical term, is to rear larvae under constant light until they pupate and then transfer the pupae to continuous darkness. About 4 days later the adult flies begin to emerge, but they do so during a particular phase of the 24 h cycle that corresponds to the period just before dawn. Since the pupae were not all at exactly the same age when they were transferred to the dark, and the rates of development and metamorphosis to the adult form vary among individuals, the adults emerge on different days – but they do so during the same limited period of the cycle. The clock effectively opens a door at one phase of the cycle and lets the metamorphosed fly step out if it is ready, then shuts the door several hours later. If maturation occurs when the door is closed, the fly has to wait. The result is a pattern of emergence as shown in Figure 7.8, At 20 °C this circadian rhythm has a mean period of 24 h 7 min in the dark.

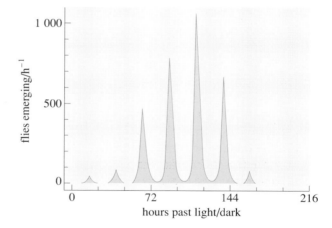

**Figure 7.8** The pattern of emergence of adult *Drosophila* from the pupal case as a function of time, showing a circadian rhythm.

One of the basic properties of circadian rhythms is that the rhythm can be shifted or reset to a new phase by external signals, light being one of the most effective. This is to be expected from the very fact that these rhythms have periods that are close to, but not identical with, 24 h, so that every day they experience a slight correction by the external day–night cycle. It was found that a brief light signal in the dark period is enough to shift the clock to a new phase, and that the size and the direction of the shift, whether forward (phase advance) or backward (phase delay) depends upon the time of the cycle. The pupae in the dark are likewise sensitive to a brief light signal which shifts the time of the eclosion rhythm, advancing or retarding it (or, for particular stimulation times, leaving it un-changed). One other property of these clocks is relevant here. When kept in continuous light, the clocks stop. For example, if *Gonyaulax* cultures are kept in constant light of intensity equal to daylight, the rhythm of luminescence and cell division disappears. Cells continue to have luminescent flashing and to divide, but these activities are no longer organized on a 24 h basis. Similarly, *Drosophila* kept in constant light throughout larval and pupal stages have no eclosion rhythm. The adults emerge as soon as they are ready, and there is no clock controlling their emergence. This shows that clocks are not *necessary* for these processes; they add a dimension of order to a system that can function without them.

Now we can consider the classic experiment by Arthur Winfree, carried out in the late 1960s at the University of Chicago, which revealed something fundamental about the dynamics of the *Drosophila* eclosion clock, and was later verified for other rhythmic systems. Winfree reasoned that biological clocks are basically limit cycles, as described in Section 7.2.1. In Figure 7.6b a particular type of limit cycle was described, the behaviour of a real pendulum, the classic clock mechanism. We are now going to extend this to a general limit cycle description, and apply a dynamic analysis to biological clocks. The reasoning is perfectly general and can be applied to a very large class of oscillating processes.

Suppose that the *Drosophila* eclosion rhythm is represented by periodic behaviour of a pair of variables, $X$ and $Y$, which together describe the dynamics of the process (like the variables $u$ and $v$ used to describe the pendulum in Section 7.2.1, but now $X$ and $Y$ represent biological variables). These variables might change rhythmically in time as shown in Figure 7.9.

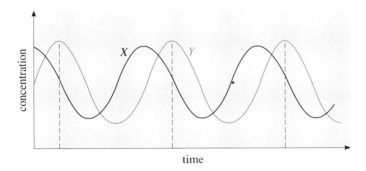

**Figure 7.9** Hypothetical rhythmic pattern of two substances with concentrations $X$ and $Y$ that are involved in generating a circadian rhythm in an organism.

Since every cycle in Figure 7.9 repeats itself, we can restrict our attention to any one, say that between the dotted lines. Now instead of representing $X$ and $Y$ as functions of time, plot them against one another. The result is a closed curve, shown in Figure 7.10 – closed because the cycle exactly repeats itself, so successive cycles are obtained by simply going round and round this cycle. Since this cycle is the stable oscillation of this system, if $X$ and $Y$ are started at any point

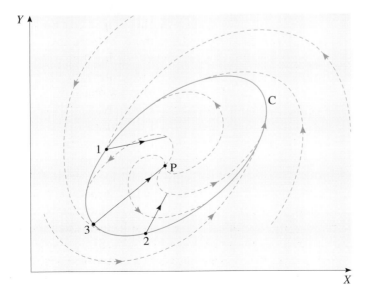

**Figure 7.10** The pattern of change of concentration of substance $Y$ in relation to $X$, showing the stable cycle as a closed curve (C) and the transients leading to the closed cycle as dashed curves. The point P within C is a stationary point where there is no change of $X$ and $Y$. 1, 2 and 3 represent different times in a stable cycle at which light flashes disturb the dynamics and change the state to the point at the end of the arrow, representing the disturbance.

off the cycle they will converge on the cycle, as shown by the dashed lines. The whole space is filled with these trajectories, so that wherever the system starts it will end up on the limit cycle, C.

The first thing we can do with this figure is to understand the dynamic origin and nature of phase-resetting by a stimulus such as light. Suppose that C represents the *Drosophila* eclosion rhythm of pupae in the dark. A flash of light given at the phase of the rhythm labelled 1 (Figure 7.10) is going to disturb the clock in some way, which means shifting it off the cycle and onto another trajectory. Suppose this is represented by the arrow from point 1. The light flash is brief (of the order of a minute or less) and then switches off, so the disturbed clock now returns to its normal dynamics – i.e. it follows the dashed orbit back to the limit cycle, rejoining it at some point. Compared with a pupa on an undisturbed cycle, this light-stimulated oscillator will be *delayed* because of the shift off the cycle and onto a trajectory whose phase is delayed relative to point 1 on the cycle. This will be registered as a phase delay in the stimulated pupae compared with a control group.

Suppose now that the light stimulus comes at the phase indicated by point 2. Again the system will be knocked off the cycle and onto another orbit, shown by the arrow. This time the disturbance takes the system *forward* in phase so that when the new trajectory rejoins the cycle it is ahead of a pupa on an undisturbed cycle. The result is a phase advance. Phase shifts at every point of the cycle can be mapped in this way so the dynamic picture of Figure 7.10 matches qualititively the pattern of phase shifts observed experimentally. Now comes the interesting prediction and experimental test.

Notice that there is a point, P, where all the spiralling trajectories meet within the limit cycle, C. This is called a **singularity** of the dynamics, a point where there is no motion and no phase to the dynamics. If the system is started at this point, it stays there until some random disturbance knocks it into a neighbouring trajectory, where it starts its journey towards the limit cycle. Whichever trajectory it happens to get kicked to will determine its phase on rejoining the cycle. In a real system, P won't be a mathematical point but a region of finite size. Winfree reasoned that, if such a region existed in the fruit-fly's eclosion clock, then a light signal of the right intensity (S*) delivered at the right time in the cycle (T*) would drive the system to this point. However, finding this critical pair of values was like looking for a needle in a hay-stack. By dint of some very ingenious experimental design and automation, Winfree succeeded in setting up a system to catch a singularity by scanning a wide range of (T, S) pairs. And he found the critical pair at T* = 6.8 h (measured from the point where the pupae are transferred from light to dark, starting the clock) and S* = 50 s of a standard blue light used in these experiments. When pupae were given this stimulus at this time, the eclosion rhythm vanished: the culture became arrhythmic, flies emerging at any time, whenever they were ready.

▷ A culture is given the critical stimulus, S*, at time T*, so the clock stops. Shortly after this, another light stimulus is given. What do you predict to be the behaviour of the pupae after the second stimulus?

▶ The second stimulus will shift them away from the point P and onto some trajectory on the other side of P from 3. So these pupae will become rhythmic again, with a phase shift determined by their position after the second stimulus.

Winfree's result and subsequent work contributed significantly to a new era of analysis of biological rhythms. The study of the properties of biological rhythms via their general dynamic properties was demonstrated to be not only possible but experimentally useful, and testable. This was soon extended to biological processes that are not only organized in time, but are patterned also in space, which have come to be known as **excitable media**. These will be examined in Section 7.3.4. For the moment we shall stick with temporal organization in *Drosophila* and look at the results of a very different approach to the study of the circadian clock – that which uses genetic analysis, for which *Drosophila* is the ideal organism.

## Clock mutants in Drosophila

The application of genetic methods of analysis to the study of biological clocks has been very instructive on two counts: it has given important insights into the molecular mechanisms involved, and also insights into the dangers of adopting particular views of how and where genes act in physiological regulation. The power of genetics is that, in principle, any protein involved in any process can be selectively changed by a mutation in the gene coding for that protein, so that a molecular dissection of the elements of the targeted process is possible. But it is necessary to devise methods of detecting defective organisms carrying mutant genes and this is where limitations can arise from assumptions about the nature of the molecular processes involved. In the 1960s, the prevailing view was that the circadian clock was so centrally involved in physiological regulation that any mutation that altered its function significantly would probably be lethal to the organism, so that all such mutants would be dead.

▷ What property of the *Drosophila* eclosion clock presented earlier in this section could be used against this view?

▶ The clock can be stopped either by exposing pupae to continuous light or by a critical stimulus without affecting the viability of the emerging fly.

▷ What is the difference between these ways of stopping the clock and a genetic mutation?

▶ The mutation alters a gene (DNA) and consequently changes a protein so that it fails to function normally. Stopping the clock by light (continuous or by a critical signal) leaves all molecules functionally normal, only the dynamics being altered.

Fortunately there are always students who ignore conventional wisdom and pursue their own deviant paths. A graduate student at the California Institute of Technology called Ronald Konopka attacked the problem of finding clock mutants by the most direct route, inducing mutations in larvae by the use of X-rays. Among the survivors he found flies whose clock was significantly different from that of normal flies. To identify these he made use of a circadian rhythm of locomotor activity that had been well characterized: adult flies are more active in the evening and early morning. Konopka found mutants in which the normal rhythm of about 24 h was altered to either shorter (16–19 h) or longer (28–30 h) periods. He called these *period* mutants (abbreviated to *per*), and he was soon able to show that these mutants all had modifications in the same gene. The eclosion clock is affected in the same way as the rhythm of locomotor activity so both processes are evidently connected to the same circadian rhythm, as previously concluded from observations that both rhythms show the same type of temperature compensation and phase-resetting behaviour. The shorter and longer period mutants (carrying *per* s and *per* l alleles) are still entrained to 24 h LD cycles, so neither photosensitivity nor the characteristic entrainment properties of those oscillators is lost as a result of the mutation. However, both temperature compensation and the phase of the entrained rhythm are altered in the mutant, pointing to these being linked together.

The initial findings of Konopka initiated a new era in the study of the circadian clock, dominated by the search for other mutants and the use of molecular techniques to identify the nature of the gene products. These will be described in detail in Book 4.

A variety of lines of evidence had pointed to the brain as the location of the circadian clock in *Drosophila*, and in insects generally. Photosensitivity involves the brain. Transplanting the brain from one moth to another is equivalent to transplanting the circadian schedule. In *Drosophila* genetic techniques made it possible to show that only brain tissue need carry the mutation for the fly to show altered circadian behaviour rhythms. These observations suggest that the product of the *per* allele should be found in the brain. And indeed this has been confirmed; the protein product of the *per* allele has been identified in the photoreceptor cells of the eye, in the optic lobes and central brain (both in glia cells and in neurons).

The range of influence of the clock on behaviour turned out to be much more extensive than expected. Normal male fruit-flies have a courtship song that consists of a repeated high-frequency wing beat that generates a tone in the same way that male grasshoppers produce their rasping note, though *Drosophila* is much quieter. It was discovered that the *per* s mutant, which shortens the circadian rhythm to 18–20 h, also shortens the rhythm of the interpulse interval of the male courtship song. The normal rhythm of notes has a period of 55 s. This is shortened to 35–40 s in the *per* s mutant, while the *per* l mutant with a lengthened circadian period (28–30 h) has a lengthened period of the courtship song rhythm of 75–90 s. Evidently the same gene product is influencing frequencies that differ by three orders of magnitude. The courtship song oscillator is located in the thorax, not the head, so the *per* allele is presumably also active in this part of the body.

How is the product of the *per* allele acting? The gene has been sequenced and the amino acid sequence of the protein product determined, revealing regions with similarity to three other proteins believed to function as nucleus-localized transcription factors (these are involved in gene regulation, as you will learn in more detail in Book 4). The concentration of the *per* protein (known as PER) oscillates with a circadian period and appears to exert a negative feedback on its own rate of production, reducing the rate of *per* transcription. PER is also predominantly localized within the nuclei of the cells in tissues where it has been identified, which is consistent with a regulatory function on circadian gene activities. It also occurs in very small quantities, as expected of a regulator of the clock.

This is where we return to an assessment of genetic approaches to physiological and related problems. The design of the genetic experiments that resulted in the discovery of the *per* alleles in *Drosophila* was such that flies were selected whose abnormalities were restricted to modifications of clock function. The consequence is that the only genes identified were those whose activity is restricted to time-keeping, without involvement in more basic metabolic activities whose disturbance could more seriously affect the development and behaviour of an organism with the mutation.

A curious terminology has arisen that distinguishes between different categories of gene, reflecting a predominant manner of conceptualizing biological process. The basic metabolic activities on which life depends are called 'housekeeping' functions (Chapter 3). No one likes housework, particularly geneticists, it seems. The genes that *regulate* housekeeping activities, and those which switch on and produce the more specialized proteins that give cells their differentiated properties (whether muscle, nerve, follicle or oocyte), have been designated 'master' and 'luxury' genes, having an altogether more elevated status. Such metaphors can actually influence the design and interpretation of genetic experiments. The entry of genetic analysis into the study of biological clocks with the discovery of the *per* alleles in *Drosophila*, and equivalent genes in the fungus *Neurospora* that will be discussed next, have resulted in 'the transformation of the entire field of rhythms from a comfortable phenomenologically-based scientific backwater into the well-recognized fast-lane research problem that it is today', according to a reviewer.

The term 'comfortable phenomenologically-based scientific backwater' dismisses decades of ingenious and painstaking research that identified and characterized circadian rhythmicity as a virtually universal property of organisms, a fundamental scientific achievement in its own right. But the lack of historical perspective revealed in the above quotation is perhaps not such a serious deficiency as the misunderstanding of scientific principles that this quotation reveals. 'Phenomenologically-based' in this context means a phenomenon that is simply observed, without an explanation of its causes. The implication is that circadian rhythms do not have an explanation at the level of dynamic properties in terms of phase, frequency, response to external stimuli, and the deduction of points of singularity such as that described in Figure 7.10, which resulted in Winfree's prediction and his remarkable result in stopping the eclosion clock.

▷ What principle of scientific explanation, described in Book 1, Chapter 1, does this violate?

▶ That different levels of organization require different explanatory rules and there is no level that is 'fundamental' as the source of causes and explanations in biology. What the reviewer is denying is that the properties of circadian rhythms require principles of explanation at their own level. His belief is that only when you get to the molecules involved do you reach a satisfactory level of explanation of the phenomenon. This is a reductionist view.

▷ What level do you need to go to in order to explain how proteins like PER assume their 3-D structure and their kinetic properties that allow them to act as control molecules by recognizing and binding to specific sites?

▶ The level of thermodynamics to explain protein folding, and charge fluctuations at recognition sites to give resonant energy interactions that result in binding.

▷ Isn't this a more 'fundamental' level of description than identifying the primary amino acid sequence of the molecule, which is what DNA sequencing studies give?

▶ Different, yes; more fundamental, no.

Why is it incorrect to claim that a knowledge of the molecules involved in a process such as a circadian rhythm gives us an explanation of the phenomenon? Because to explain the phenomenon we need to put the molecular pieces together into a dynamic description that involves concepts such as period, phase, phase-resetting, temperature compensation, etc., that do not belong to the molecules themselves but to the dynamic system to whose operation they contribute. We can describe the behaviour of this system, the circadian oscillator, in terms of a dynamic model such as that presented in Figure 7.10, which can be used to explain and study properties such as phase-resetting and clock-stopping. There must be some consistency between the behaviour of the molecules involved and descriptive variables such as the concentrations of $X$ and $Y$ in Figure 7.9. However, it is not necessary to know all the molecules involved and their properties in detail to understand the circadian oscillator dynamically.

Scientific explanations involve abstracting out the essential properties of the system and relating these to the activities of constituent elements. Without the causal analysis of clock dynamics that focused on certain essential features of these oscillators, one would not know what properties to seek at the molecular level, such as those that could be involved in temperature compensation. So the integrated dynamics of the system and a knowledge of the properties of its constituent elements that are involved in producing higher-level emergent behaviour are equally important areas of analysis, neither taking priority over the other except by the dictates of fashion or prejudice. (You may wish to look back at the discussion of emergent properties in Chapter 1.) There is, however, a prevalent, and unfortunate, prejudice that is associated with molecular and genetic reductionism in biology that reflects a failure to recognize the importance of different levels of explanation. The other prejudice, that 'interesting' genes belong to higher-order 'master' or 'luxury' classes, influences experimental design in the ways that are evident in the identification of the *per* alleles in *Drosophila*. Use a selection procedure that identifies a control or regulatory gene and you won't find one that is involved in the dynamic level where circadian rhythms emerge from basic physiological and biochemical organization. The *per* genes are certainly interesting and important but they don't reveal the dynamic nature of the clock. Getting closer to this requires a different way of thinking about the problem, and consequently a different experimental design.

### 7.2.4   The rhythmic mould

The bread mould *Neurospora crassa* has, like *Drosophila*, a celebrated genetic pedigree. It was used by George Beadle and Edward Lawrie Tatum in the 1940s to establish the discipline of biochemical genetics, which identified the relationship between genes and enzymes involved in metabolic pathways. A circadian rhythm in *Neurospora* was reported in 1959 by a group working at Princeton University led by two biological clock pioneers, Colin Pittendrigh and Victor Bruce. The fungus grows as a tangled mat of filaments called hyphae that spread on a nutrient surface, growing as a disc with a roughly circular perimeter. Towards the end of the dark period and into the early morning there is a change in the growth pattern: as the hyphae grow along the surface, they also produce aerial shoots that differentiate into spores called conidia. These spores are released when the aerial hyphae dry out and they are then dispersed by wind or mechanical disturbance. If they land on a nutrient source, they germinate and start a new mould. The spores are tiny and remain viable for several years, so they readily become distributed and lie in wait in the most unlikely place for food you happen to leave around.

The circadian rhythm of conidiation in *Neurospora* is an example of an oscillation that results in a spatial pattern: a mould growing on a surface generates concentric rings of aerial hyphae that are clearly distinguishable from the vegetative hyphae, as shown in Figure 7.11. The growing region of the mould is like a wave front that propagates over the nutrient medium, changing its state with a 24 h cycle when exposed to external light–dark signals. In the dark at 25 °C, the period of the rhythm is 21.6 h. Konopka's success in finding clock mutants in *Drosophila* prompted another young researcher, Mark Feldman, to try his luck on *Neurospora* using the same selection methods as Konopka: he looked for fungi that were normal in all properties except the period of the clock. It was several years before he succeeded in finding them. Eventually they turned up and were called *frequency* mutations (*fr*), all mapping to the same genetic locus. Their effects were like those of the *per* gene mutants in *Drosophila*: they altered the natural period (or frequency) of the circadian rhythm, the shorter periods being about 16.5 h and the longer ones up to 29 h – also as in *Drosophila*.

These mutants have also revealed significant properties of the oscillator, such as a decrease in the sensitivity of the clock to phase-resetting signals as the period increases from 16.5 to 29 h, and differences in temperature compensation in the different alleles. Such results have given rise to specific dynamic models that can explain these aspects of clock behaviour, and make useful predictions about others. But the basic experimental design of looking for mutants whose action is exclusively on clock period makes it difficult to get at the metabolic origins of the oscillation itself.

An altered perspective comes from the work of a group in San Diego in the 1980s, led by Stuart Brody. The strategy was to examine *Neurospora* mutants with known biochemical defects that also affected some aspect of clock function. This approach accepts the conclusion of earlier studies that the circadian oscillator is probably intimately connected with basic metabolic activities ('housekeeping functions'). Therefore, mutations in genes whose products act at the heart of the circadian rhythm would affect not only aspects of clock function but also fundamental processes such as growth. A member of this group, Patricia Lakin-Thomas, examined growth mutants for altered circadian behaviour and discovered one with extremely interesting properties. The mutant strain was called *chol*-1 because it requires choline, a metabolic precursor of phosphatidyl choline (a key constituent of membrane phospholipids, see Book 1, Chapter 1), for normal growth.

At choline levels that result in growth rates above 1 mm h$^{-1}$, the circadian rhythm is normal in all respects. As choline is decreased and growth rates fall below this value, there continues to be a regular rhythm of conidiation but its period increases from the normal 21.6 h to more than 60 h when the growth rate has decreased to 0.6 mm h$^{-1}$. These rhythms with different periods continue to have clock properties: they entrain to light–dark cycles, but the LD regime must match the intrinsic rhythm quite closely. 'D' now means 'dark', rather than dim light, as a mould such as *Neurospora* does not have photosynthetic activity and lives on nutrients in its environment. For example, mutant cultures supplemented with choline at a concentration of 10 µmol l$^{-1}$ have an intrinsic period (in DD) of 23.8 h and can be entrained to a 24 h L : D cycle (e.g. L : D = 12 : 12). These cultures do not entrain to L : D = 18 : 18 or 24 : 24 (corresponding to 36 and 48 h 'days'). Cultures on 0.5 µmol l$^{-1}$ choline have intrinsic periods of 35.4 h and will entrain to L : D = 18 : 18, while cultures with intrinsic periods of 46.6 h will entrain to L : D = 24 : 24, and so on. This makes it likely that the oscillatory mechanism itself is being affected by the mutation. What could this mechanism be?

The mutant is known to be defective in the production of phosphatidyl choline. This is a membrane phospholipid that is directly involved in the regulation of

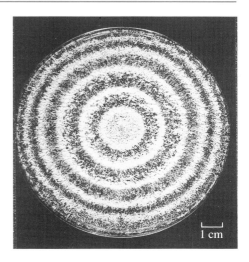

**Figure 7.11** The spatial pattern of growth rings of the bread mould *Neurospora crassa*, showing a circadian rhytym of spore formation (dark rings).

metabolic functions through the action of its fatty acid containing part, the second messenger diacylglycerol (DAG). This membrane-localized molecule influences many cellular processes via its effect on protein kinase C (as discussed in Chapter 3). Protein kinase C, among other activities, modulates ionic fluxes across membranes, which in turn influence metabolic activities. The *chol-1* mutant indicates that the circadian oscillator involves regulatory activities in the membrane that are coupled directly to the basic metabolic machinery of the cell. A similar conclusion that membrane-localized regulation of metabolic processes was involved in clock function had been deduced from less direct evidence in the 1970s and from previous work of the San Diego group on other lipid metabolism mutants that affected the clock. Another attractive feature of a membrane model of the oscillator is that the lipid composition of membranes varies with temperature, so that their fluidity remains roughly constant irrespective of ambient temperature over ranges of 10–40 °C. This could well be involved in the temperature-compensation mechanism of the circadian clock – another suggestion that came from earlier work.

What is now being achieved, therefore, is a synthesis of ideas from different levels of analysis of the circadian oscillator into an integrated approach that makes equal use of a broad range of concepts and techniques. This is paying off in opening a window onto some molecular elements of the circadian clock and how these may be integrated into basic cell functions. The ramifications of this line of study will be very extensive and it will be easy to get lost in the molecular abyss. What is required in order to proceed effectively and efficiently is a clear understanding of the essential properties of the oscillator that need to be explained, the use of molecular insights to identify how these may arise from particular properties of proteins, and their integration into a dynamic model that explains the observed behaviour.

### Summary of Sections 7.2.3 and 7.2.4

The circadian rhythm of eclosion in *Drosophila* was used to demonstrate a fundamental property: the biological clock can be stopped by a critical stimulus given at a specific phase of the rhythm, involving no damage to any of the clock components. This dynamic property can be explained by a basic feature of a general class of oscillators, which also indicates how biological clocks are advanced or delayed by an environmental signal such as light.

Mutations in a gene called *period* (*per*) affect the free-running rhythm of the clock, some alleles (*per*$^s$) shortening the period (minimum 18 h) and others (*per*$^1$) lengthening it (maximum 30 h). This gene is active in the nuclei of cells in the brain, where the clock controlling behavioural rhythms such as locomotor activity is located, and it appears to act as a regulator of the clock period. It is sometimes claimed that a detailed knowledge of the genes involved in clock function will provide a complete explanation of circadian rhythms, but to achieve this it is necessary to put the pieces together into a model that provides an integrated dynamic description of the properties of the biological clock.

The bread mould *Neurospora* has proved to be useful in discovering more of the pieces involved at basic levels of clock function. A gene called *frequency* (*fr*) has mutant alleles that alter the circadian period of growth in ways similar to alleles of the *per* gene in *Drosophila*. In addition, there are mutations of genes involved in lipid metabolism that have dramatic effects on the rhythm, extending it to more than 60 h. These point to metabolic processes that are linked to membrane properties as basic components of circadian oscillation.

## 7.3   Making waves: coupling time and space in dynamic patterns

### 7.3.1   Introduction

The concentric circles of Figure 7.11 show what can happen when a periodicity in time is combined with spatial movement, in this case the advancing front of a *Neurospora* hyphal mat growing over a surface at about $1 \, \text{mm h}^{-1}$. In the dark at $22 \, °\text{C}$, the rhythm of conidiation leaves its trace as a pattern of aerial hyphae produced every $21.6 \, \text{h}$, the rings being about $2.16 \, \text{cm}$ apart. This is an example of how an advancing wave in a system that is oscillating in time can generate a spatial pattern. Is the whole mould oscillating, or just the growing tip? The way to find out is to cut out a bit of the non-growing part, in the central mass of the mould, and to put it on fresh nutrient. It starts to grow from the boundaries and produces the same periodic pattern as the original mould. Furthermore, it keeps the same time as the original, producing conidia at the same phase of the $21.6 \, \text{h}$ cycle. And it can be phase-shifted by light stimuli in the same way as its 'parent'. So the oscillation is going on throughout the mould, but this is made visible only at the growing front where a change of growth characteristics occurs and reproductive conidia are generated.

It is characteristic of organisms with rather rigid walls, such as fungi and plants, that change of form occurs only in association with change in the pattern of wall growth. Where there is no growth, there is nothing to reveal the physiological rhythm, which nevertheless continues. So *Neurospora* is a system in which every part is changing rhythmically with the same period, and any part that is isolated from the rest just goes on with its rhythmic pattern. What will now be investigated is a somewhat different type of system in which the parts can respond to and transmit a signal so that spatial patterns can arise, but the parts are not necessarily rhythmic themselves. Many biological tissues show this type of property, which is called **excitability**. One of the most familiar is the heart.

### 7.3.2   Heart dynamics

The standard, time-honoured ways of tuning in to the heart are by using a stethoscope to monitor the sound waves caused by the powerful contractions of this unique muscular organ, or to record the change of electrical potential at the surface of the body by electrocardiography. A typical electrocardiogram (ECG) trace for a single cycle of the heart (about 1 min) is shown in Figure 7.12. Height represents electrical potential. The P wave corresponds to the initiation of the heart beat, the QRS complex reflects the main heart contraction, and the T wave shows the electrical recovery of the heart tissue. This basic pattern can be detected on the skin anywhere in the body: the heart beat is a galvanizing event.

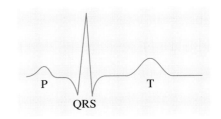

**Figure 7.12**   A normal electrocardiogram trace for a single cycle of the heart (about 1 s).

The spatial characteristics of the heart beat are shown diagrammatically in Figure 7.13. Electrical depolarization of heart tissue begins at the sinoatrial node (SAN) in the right atrium. This node is made of specialized heart cells that collectively have a rhythm of depolarization. This electrical signal propagates through the muscle tissue of the atrium (AM), accompanied by muscular contraction, and arrives at the atrioventricular node (AVN), which responds by depolarization. Specialized conducting fibres (Purkinje fibres, PF), spreading from the AVN to the muscle tissue of the ventricles (VM), initiate the main pumping action of the heart in these chambers, accounting for 80% of the power in each stroke. These tissues then recover in preparation for the next cycle. This recovery includes the restoration of electrical polarization across the membranes of the heart cells which, like nerve cells, are about 90 mV more negative in electrical potential inside than outside. The same ionic mechanisms of channel opening in membranes following a critical depolarization provide the basis of electrical excitability in heart cells as in neurons (see Chapter 2). And, as expected, they have the same type of refractory state after depolarization which lasts for 100–200 ms, depending on heart rate, during which the cells will not respond to an electrical stimulus. This prevents backfiring. The relatively flat, quiet interval following the T wave in Figure 7.12 is when the atria of the heart are filling with blood from the body (right atrium) and from the lungs (left atrium) until the SAN initiates another cycle.

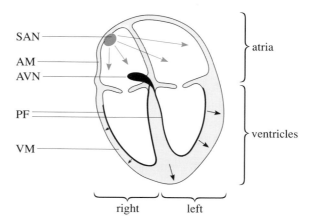

**Figure 7.13** Pathways of propagation of the action potential in the heart, from SAN to VM, showing right and left sides of the heart.

Whereas the normal, healthy heart appears to be beating in a very regular rhythm under conditions of rest, recent studies carried out at the Beth-Israel Hospital in Boston, Mass., by Ary Goldberger and his colleagues have shown that its dynamic behaviour is complex and subtle. Plotting heart rate in beats per minute against time gives a result of the type shown in Figure 7.14a. This reveals that the interval between beats is not actually regular, but varies in what appears at first sight to be a random, fluctuating pattern around a mean value of 80 beats $min^{-1}$ for much of this recording. However, a closer analysis of this pattern suggests that it is not random but chaotic, as shown in Figure 7.14b, in which the rate at different times is plotted against the rate 12 s before. (This kind of graph comparing two different rates is called a Poincaré map.) If the variations were purely random, there would be no pattern at all in this type of plot, since there would be no correlation between the rate at one time and the rate 12 s earlier. However, there is some correlation which shows up as the pattern of Figure 7.14b. This strange mixture of irregularity and order is a characteristic of systems that are technically described as **chaotic**. Does this chaotic dynamic of the heart rate have any significance? It appears so.

Figure 7.15 shows the heart rate and Poincaré map for a subject at risk of cardiac arrest who died 8 days after the electrocardiogram was made from which the plots were constructed. Instead of a chaotic pattern of heart rate variation, this individual showed a very regular change of rate, with a dominant frequency and a highly localized Poincaré map. Evidently an appropriate mixture of regularity and chaos is the healthy mode of physiological function, and too much regularity is diagnostic of potential danger. Why this should be so is not yet well understood, but chaotic behaviour as a component of physiological dynamics is becoming widely recognized as the normal, indeed the healthy, condition.

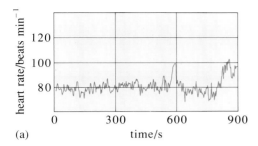

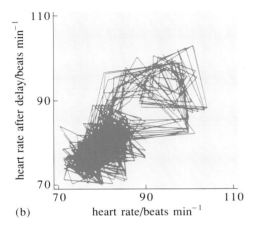

**Figure 7.14** (a) Change in time of heart rate in the normal heart, showing a highly irregular pattern. (b) The Poincaré map of the heart rate shown in (a), plotting beats per minute at time $t$ against the value at time $t - \tau$, where $\tau$ is a small delay.

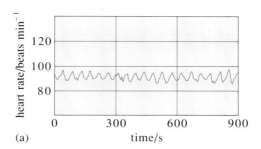

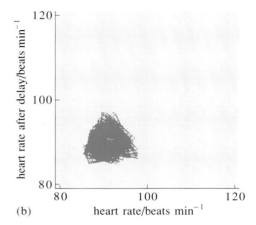

**Figure 7.15** (a) Change in time of heart rate in a person at risk of heart failure, showing a highly rhythmic variation. (b) The Poincaré map of the heart rate in (a), showing the very regular pattern of the rhythmic variation. Compare this with that of a healthy subject, in Figure 7.14b.

▷ Try the following experiment. Take your pulse while sitting quietly and feel the regularity of the beat. Then take a deep breath and hold it, all the time following your pulse. What happens to the rate? After holding your breath for 10 s or so, exhale strongly and see if you detect a change in rate. Repeat this a number of times to see if you get a consistent result, but don't overdo it. What conclusions do you draw?

▶ In general, the heart rate increases slightly with both a deep inhalation and a strong exhalation. The heart is normally responsive to a great variety of physiological changes that have different frequencies, as we have seen, so it reflects the dynamic modes of the body. A healthy heart is a mirror of the whole body, whose dynamic coherence arises from a rich spectrum of interacting frequencies.

The heart is evidently a highly robust, durable, and reliable pump. It has a number of back-up mechanisms and modulators that reinforce its intrinsic reliability and its responsiveness to the needs of the body. The AVN, connected to the Purkinje fibres, can act as an autonomous pacemaker and initiate the major ventricular contractions if for some reason the signal from the SAN fails, though clearly the pumping action of the ventricles will be less effective if the atria have not contracted first and emptied their contents into the ventricles. Heart tissue is also intrinsically excitable in response to stretching – like any other muscle tissue, as revealed in the knee-jerk reflex. (Stimulation of stretch receptors in any muscle leads to contraction of that muscle.) So simply filling the chambers with blood and extending the tissue can initiate a contraction. The whole system is constructed on the basis of properties that are natural and intrinsic to elastic, muscular tissue that can operate independently of any external nerve supply. However, sensitive coordination of heart activity with body requirements is enhanced by a network of nerve fibres that infiltrate all parts of the heart, but particularly the pacemaker nodes and the specialized conducting tissues. These belong to what are known as the sympathetic and the parasympathetic nervous systems, which record states of stress and exert a diversity of modulating and coordinating effects on the heart by such influences as shortening the refractory period and accelerating repolarization. These nervous systems are not normally subject to voluntary control, but meditational and relaxation techniques can be used to influence their activities so that variables such as heart rate and blood pressure are altered.

Despite all of these aspects of activity that make the heart such a reliable and responsive organ, it can fail in the most dramatic and unexpected way. A healthy, functioning heart can fail in its pumping activity as a result of small incidental stimuli. The phenomenon known as sudden death due to cardiac arrest claims about 25 000 victims per year in the United Kingdom. While post-mortem autopsies often reveal that such cessation of the heart is correlated with slight damage to heart tissue, often caused by ischaemia (inadequate blood supply to the heart tissue itself due to some circulatory occlusion such as a thrombus or thickening of the walls of the coronary arteries), there are many instances where the heart tissue appears to be perfectly healthy. It seems that the heart has intrinsic dynamic properties that put it at risk of malfunction independently of damage. Understanding this takes us into the dynamics of excitable media and a spectrum of disorders that are becoming known as dynamic diseases. Organisms as homeodynamic systems benefit from the rich range of emergent properties that arise from their complex dynamic organization, but not all these are appropriate for physiological function.

### 7.3.3   Cardiac arrhythmias

It used to be believed that sudden cardiac death was due to an abrupt cessation of activity in the heart. This would then be like the stopping of the eclosion rhythm in *Drosophila* by a stimulus of just the right intensity and duration at a critical phase in the circadian rhythm, as described in Section 7.2.3. However, there is an important difference between these two phenomena, apart from the fact that the *Drosophila* seem to survive perfectly well without the eclosion clock. Over a century ago, in 1888, the cardiologist J. A. MacWilliam described cardiac death as dynamic disorganization rather than cessation: 'the cardiac pump is thrown out of gear, and the last of its vital energy is dissipated in a violent and prolonged turmoil of fruitless activity in the ventricular wall'. This condition of uncoordinated contractions is now called **fibrillation**, a kind of repetitive quivering of the heart muscle as waves of ineffective contraction pass over it. It can be observed clinically in the atria, where fibrillation is not lethal, but once the ventricles start to fibrillate the body is in danger unless the condition is quickly corrected. What initiates this condition?

Cardiologists working in Maastricht, The Netherlands, reported the following case in 1972. A 14 year-old girl suddenly lost consciousness when wakened one night by a thunderclap. This suggested anoxia due to circulatory failure, but the condition was transient and the girl spontaneously regained consciousness without damage. However, she would then often faint when startled by her alarm clock in the morning, recovering one or two minutes later. The cardiologists established that this was indeed due to ventricular fibrillation, as shown in the electrocardiogram of Figure 7.16.

*Figure 7.16* The initiation of ventricular fibrillation in a person's heart (the high-frequency, regular rhythm at the right) that started soon after an alarm went off.

A common cause of such behaviour is the neural input to the heart along the pathways of the sympathetic system, as previously described. The sympathetic nerves ramify across the heart muscle and activate it when the nerve endings release noradrenalin, depolarizing the cardiac membrane via the receptors that receive the chemical signal. These are called β-receptors and they can be plugged by an artificial β-blocker molecule which is usually designed as a non-functional and non-removable chemical analogue of noradrenalin (e.g. propanolol). So the girl was given this β-blocker and her condition improved. After a few years she met a boyfriend who did not believe in drugs and she was persuaded to discontinue the propanolol. She was found dead in bed fourteen days later. Fortunately, affairs of the heart rarely have such direct physical consequences as in this tragic case.

The pattern of high-frequency ventricular contractions in the latter part of the ECG in Figure 7.16 is an example of tachycardia (fast heartbeat), which are the commonest types of cardiac arrhythmias, but bradycardias (slow heartbeat) also occur. Although these are described as arrhythmias because the normal contraction and relaxation cycles fail to occur, they are more accurately described as dysrhythmias, abnormal and disorganized rhythms. What properties of the heart tissue itself make it vulnerable to such dynamic behaviour?

### Stopping the clock

The similarities between the heart beat and the circadian clock as oscillators, despite their great difference in frequency, suggest that Winfree's dramatic result of stopping the circadian eclosion rhythm of *Drosophila* (Section 7.2.3) with a critically-timed impulse (light) may well have an analogue in the heart. Notice that this reasoning depends upon general, typical properties of oscillators (what mathematicians call their generic properties) and is not related to the properties of the components that make up the oscillators. What is important is the way these are organized as a dynamical system, not their composition. So we are now focusing on properties characteristic of a particular level of dynamic organization that cannot be understood in terms of the properties of their parts – i.e., these are emergent properties.

The type of experiment that Winfree carried out on *Drosophila* larvae has in fact been used for many years in the study of heart dynamics. Instead of a light signal, the appropriate stimulus to the heart is a small electrical pulse delivered at different times in the rhythmic cycle. Experiments of this type were carried out on the rabbit heart by G. R. Mines, working at McGill University in Montreal, Canada, in the period 1910–1914. He applied a small electrical stimulus to the apex of the ventricle at different phases of the cycle. Most of these advanced, delayed, or left unchanged the phase of the next cycle, just as light signals produce

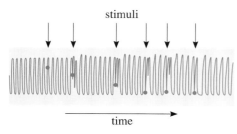

**Figure 7.17** Small electrical impulses (represented by arrows) delivered to a healthy rabbit heart can cause brief episodes of fibrillation (high-frequency cycles) but the normal rhythm usually returns.

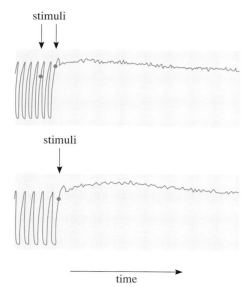

**Figure 7.18** A small electrical stimulus delivered to a healthy heart can cause cessation of normal heart beat if it arrives at a critical time in the cycle.

characteristic phase-resetting responses in a circadian clock (Section 7.2.3), but, after a brief disturbance, contractions of normal amplitude resumed. Such responses are shown in Figure 7.17, which is from a paper by Mines published in 1914. However, he discovered that there is a critical phase of the heart cycle where the stimulus produces the dramatic effect shown in Figure 7.18: the ventricular tissue goes into prolonged fibrillation. These experiments have since been extensively repeated and confirmed. This is the analogue of the circadian clock-stopping experiment, though discovered half a century earlier. But what now needs to be explained is the phenomenon of fibrillation itself, which appears to be distinctive to the heart. The heart does not stop its activity but changes its dynamic mode, from a coherent propagating wave of contraction that sweeps from atria to ventricles (Figure 7.12), into a self-sustaining high frequency quivering in the ventricles.

Isolated bits of heart tissue do beat spontaneously, though at rates much slower than the normal heart beat. The normal rhythm comes from the activity of the dominant pacemaker, the SAN (Figure 7.13). So how does the ventricular tissue become the source of high-frequency propagating waves of contraction that fail to pump blood and result in anoxia which, if prolonged, leads to death? A clue to this behaviour comes from autopsies performed on sudden death victims whose hearts showed small regions of damaged tissue in the ventricles. Such islands of damage, called infarcts, can arise from local failure of blood supply to the tissue due to coronary occlusion, a block in the circulation arising from any one of the many factors initiating coronary heart disease. There can also be temporary local occlusions ('Prinzmetal's angina' or 'transient myocardial ischaemia', both due to spasm in a coronary artery) that temporarily makes the muscle more susceptible to arrhythmias. But neither of these observations explains how tissue that is not spontaneously rhythmic turns into a high-frequency pacemaker. This is where an analysis based solely on time, as carried out in Figure 7.10 (Section 7.2.3), is inadequate and the spatial dimension has to be added.

Look again at Figure 7.13. Consider first the effect of an infarct or island of damaged tissue in an otherwise healthy ventricle. A wave of depolarization spreads across the ventricle, starting with a near-instantaneous initiation by the Purkinje fibres that ramify over the interior surface of the ventricle and propagating to its outer surface. An infarct in the tissue will alter the normal pattern of wave propagation because the damaged cells will either be unable to respond, having lost their excitability, or their response characteristics will be altered. Such a spatial inhomogeneity is enough to cause a local fibrillation because a wave can be set up that propagates round and round the infarct. From the beginning of Section 7.3.2 you should remember that the refractory period of cardiac tissue, the period of time during which it is recovering from a previous depolarization and is unresponsive to a depolarizing stimulus, is 100–200 ms.

▷ What happens when two waves of depolarization, initiated from two different sites on a heart by electrical stimuli, meet?

▶ They annihilate one another because the tissue immediately behind them is refractory. This means that the waves cannot propagate through each other and into tissue that has just been depolarized.

A small infarct will not result in a fibrillation because the time for a depolarizing wave to propagate around it will be shorter than the refractory period, so the wave will run into tissue in an unresponsive state and so will stop. But if the infarct is large enough, a wave of depolarization can propagate around it indefinitely, the highest frequency being set by the refractory period, say one cycle every 200 ms, or 5 Hz.

Figure 7.19 shows an electrocardiogram of a patient known to be at risk who was wearing an ECG recorder at the time of his sudden death.

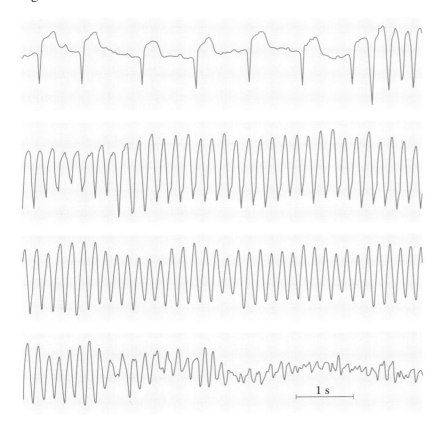

Figure 7.19 The electrocardiogram of a person at risk of a heart attack, showing the initiation of fibrillation towards the end of the top trace and death due to anoxia in the bottom trace, where rhythmic contractions cease.

▷ What is the mean frequency of the ventricular fibrillation during tachycardia (the regular high-frequency oscillations)?

▶ About 5 Hz.

▷ What appears to have initiated the onset of this tachycardia episode?

▶ The premature contraction that occurred after the seventh normal beat in the upper trace

### 7.3.4 Propagating waves in excitable media

The term 'excitability' is usually used to describe the particular type of responsiveness shown by nerve and muscle cells to an electrical depolarizing stimulus. When this stimulus exceeds a critical value, ion channels in the membrane open and the voltage difference across the membrane changes, as described in Chapter 2. The depolarizing wave spreads across the tissue of the heart because the cells are coupled to one another, the tissue behaving as a continuous medium rather than as a collection of individual cells. So we get the concept of an excitable medium in which electrical activity and (in the case of the heart) mechanical contraction waves generate patterns in space and in time. The normal heart beat is one of these patterns, stabilized by the reinforcers of the normal contraction wave such as the activity of the specialized pacemaker and conducting fibres described in Section 7.3.2. Clearly the heart is designed to function in this dominant mode. But the frequency with which fibrillation occurs, even in undamaged tissue, suggests that this is another typical or generic dynamic mode readily available to excitable media.

The first direct experimental evidence for this came from studies reported in 1973 by Dutch cardiologists. They demonstrated that a single, weak electrical stimulus to the atrium of a rabbit heart just before the normal pulse from the SAN initiated fibrillation with a characteristic pattern of electrical activity: a wave of depolarization circulated around atrial tissue that was excitable but not spontaneously rhythmic at high frequency, and was perfectly healthy. So no infarct or any kind of unusual state in an island of tissue was necessary to produce fibrillation; any region of the atrium stimulated at the critical moment in its cycle (already identified by Mines; see Figure 7.18) could sustain fibrillation. This finding led to a conjecture about the nature of this activity which came from studies of other excitable media, quite different in their nature from heart tissue, such as particular types of chemical reaction and processes occurring during the development of a particular type of organism, the cellular slime mould. In these systems, which will be described in Section 7.3.5, spiral waves of activity were observed propagating from a central region in which an activity wave propagated round and round in a circle. This 'rotor' could arise anywhere in the medium. So it appeared that the rotor that drives heart fibrillations might be of the same type, and the propagating waves might be spirals. This has now been confirmed by detailed studies of isolated, healthy heart tissue, which can sustain precisely this kind of dynamical wave. The observations are as follows.

Small squares (about $2\,\text{cm} \times 2\,\text{cm}$) of normal ventricular muscle from a sheep heart were isolated in oxygenated physiological saline (a solution whose salt concentrations are the same as those in blood plasma) with glucose and maintained at $37\,°\text{C}$. The tissue was placed on a square array of electrodes used to stimulate it, and a voltage-sensitive dye was introduced into the cells so that the electrical potential across the membranes of the cells could be recorded visually. A single stimulus from any of the four electrodes defining the sides of the square of tissue resulted in a single electrical wave that propagated across the tissue. Contraction was prevented by an electromechanical uncoupler (diacetyl monoxime), which did not interfere with electrical activity but prevented mechanical distortion of the tissue.

The critical experiment involved two successive single stimuli from line electrodes on adjacent sides of the square (i.e. at 90° to each other), at an interval such that cells were just recovering from the first wave when the second arrived. Because of the 90° angle of the depolarization waves initiated by the line electrodes, the second stimulus arrived when cells stimulated by the first wave were at different stages of recovery. The result was that a wave began to propagate in a curve within the square of tissue as shown in Figure 7.20, at 0 ms. The first electrode to fire was the left-hand vertical; the second was the bottom horizontal. The depolarization wave from the second electrode, initially a straight line, becomes distorted because of the different recovery states of the tissue after the first wave. Successive 15 ms frames show the subsequent behaviour of one square.

The reason for the non-symmetric pattern in Figure 7.20 is that the elongated cardiac cells are orientated horizontally in the preparation so that they conduct more rapidly in the horizontal than in the vertical direction. Following the frames successively in time, it is clear that the repeat of the state at 30 ms occurs at 225 ms,

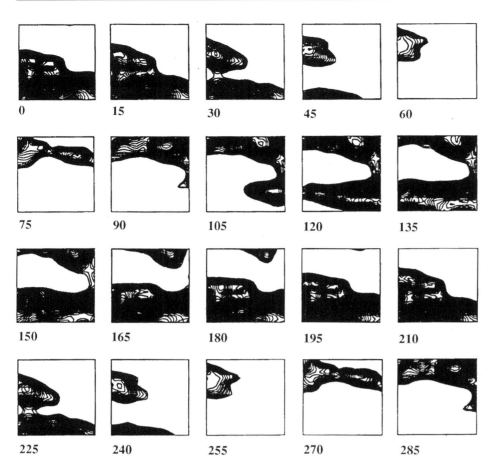

**Figure 7.20** The pattern of electrical activity in one square of sheep heart tissue, at 15 ms intervals, showing the occurrence of a re-entrant cycle characteristic of fibrillation. The intensity of the dye records the degree of depolarization.

that at 45 ms. is repeated at 240 ms, and so on. Clearly this tissue is sustaining a cyclic repetition of a depolarization wave that travels in a closed circuit around the centre of the tissue. This cyclic repetition is called **re-entry behaviour**. Tissue that is not intrinsically rhythmic has become a high-frequency pacemaker.

▷ What is the cycle time (period) of this wave?

▶ About 195 ms.

▷ How does this compare with the period of the tachycardia rhythm in Figure 7.19?

▶ It is virtually the same: 5 Hz = 1 cycle per 200 ms.

These results on cardiac tissue establish a firm foundation for understanding tachycardia and sudden death. Because the functioning heart is a very much more complex system than the experimental models, there is a great range of possible dynamic behaviours that can accompany fibrillation. A description of these possibilities by W. E. Garrey in 1914 shows the extent to which subsequent results were anticipated by his astute observations:

> Impulses can spread in any and all directions, their progress being limited only by the prexistence or development of localized blocks within the tissue mass. Such blocks divert the impulse into other and more circuitous paths, and the area so blocked off can participate in contraction only when an impulse which has passed the other parts of the ventricle approaches it from another direction; this area in turn becomes a centre from which the progress of contraction is

continued, to be in turn diverted by other blocks. The existence of such blocks, and especially of blocks of transitory character and shifting location, has been noted in the experiments detailed above. These conditions make possible the propagation of the contraction wave in a series of ring-like circuits of shifting location and multiple complexity. It is in these circus [cyclic] contractions, determined by the presence of blocks, that we see the essential phenomena of fibrillation.

Although fibrillation is often associated with blocks due to infarcts or other causes of local damage or enhanced sensitivity, the most recent work has shown that this is not necessary. Perfectly healthy tissue, if stimulated at critical times by spatially patterned excitations such as those that can arise from sympathetic nerve activity, can be plunged into one of its natural dynamic modes that leads to disaster, as in the unfortunate case of the Dutch girl whose over-active sympathetic system seems to have been the cause of her sudden death from cardiac fibrillation. So the heart, reliable pump that it is, also has its dynamic black hole. However, what is a black hole for an excitable medium operating in one dynamic mode is the natural operating state for another, as will now become evident.

### Summary of Sections 7.3.1–7.3.4

When the spatial dimension is added to a rhythmic process the result is a propagating wave as in the growth pattern of *Neurospora*, or periodically repeating waves such as occur in the normal activity of the heart. The heart has specialized structures that reinforce the one-way propagation of the contraction wave from atria to ventricles that pump the blood. However, there is a significant frequency of cardiac arrest in humans due to the onset of a disorganized contraction pattern known as fibrillation. This is often due to islands of damaged tissue is the heart (infarcts) arising from reduced blood supply to the tissue, but even perfectly healthy hearts can start to fibrillate. Models of the heart as an excitable medium, and experimental studies on animal heart tissue, have shown that fibrillation in a healthy heart can arise from an electrical stimulus whose timing and spatial location are such as to initiate a local wave of excitation that propagates in a closed cycle. The result is high-freqency, ineffective contractions, characteristic of fibrillation.

### 7.3.5 Aggregation in cellular slime moulds

In 1965, Günther Gerisch in Germany observed a striking phenomenon during the aggregation phase in the life cycle of a little organism called the cellular slime mould, *Dictyostelium discoideum* (see Figure 1.1 in Chapter 1).

This species has two distinct forms: one unicellular, in which individual amoebae crawl about in decomposing vegetation, devouring bacteria, growing and dividing; and another multicellular, in which several thousand amoebae unite to produce a structure known as a fruiting body (Figure 7.21).
The transition from one form to the other is known as the aggregation phase, during which scattered amoebae signal to one another and gather together to produce the developing multicellular organism. It was during this process that Gerisch, working with laboratory cultures in which the amoebae are placed on an agar surface in Petri dishes, observed patterns of cell movement of the type shown in Figure 7.22. Most of these are spirals whose centre is a rotor, a wave of activity that rotates in a circle, just like that in Figure 7.20. What is happening at the

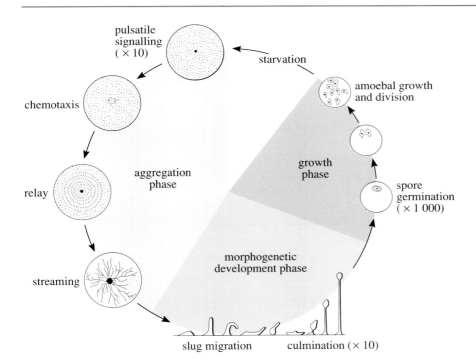

**Figure 7.21** The life cycle of the slime mould, *Dictyostelium discoideum*, showing the various stages and the spatial patterns produced during aggregation.

**Figure 7.22** Photograph of the circular and spiral patterns produced during aggregation phase of the cellular slime mould, *Dictyostelium discoideum*.

molecular level between cells in Figure 7.22 is quite different from the intercellular events occurring in Figure 7.20, although they result in the same dynamic pattern.

The amoebae are giving out pulses of the chemical cAMP, already identified as a major component of intracellular second messenger systems (Chapter 3). The pulses are released in response to the same signal from adjacent amoebae, the chemical diffusing through the agar and stimulating any amoeba that encounters a concentration of cAMP above a critical value. This is the positive or amplifying aspect of the signalling process, analogous to the transmission of a nerve impulse across the heart tissue discussed in Section 7.3.2. In addition to releasing a pulse of cAMP in response to the diffusing signal, the stimulated amoebae also move towards the source of the signal by sending out pseudopods in this direction. It is these waves of movement that Gerisch observed, which show up as the light patterns in Figure 7.22. Notice that not all the waves are spirals: several are concentric circles, a pattern to be discussed shortly. What is happening in both patterns is a one-way propagation of the cAMP pulse signal across the lawn of amoebae, resulting in ordered patterns of cell movement towards the origin of the propagating wave which in the case of a spiral is a rotating cycle of the same kind as that which can occur in heart tissue, though at a very different frequency. The rate of signal propagation in the slime mould is about 4–6 mm min$^{-1}$ compared with a velocity of 5–10 cm s$^{-1}$ in cardiac tissue, while the time interval between waves in the slime mould is several minutes compared with a second or so for the heart. However, the dynamic properties of the two systems have basic similarities.

▷ From the pattern where waves from two different centres meet in Figure 7.22, what deduction can you draw about the capacity of amoebae to respond to a second signal soon after they have responded to a first one?

▶ Since the waves are annihilated at such an interface between the patterns, set up by two centres, they clearly cannot propagate through one another, so it can be deduced that amoebae must become unresponsive for a period of time after being stimulated; that is to say, they have a refractory period.

▷ How could you estimate the refractory period from the dynamics of rotating spiral waves?

▶ If you measure the time interval between two successive spiral waves passing a fixed point on the surface, which is also the time for one complete cycle of the rotating centre, this will give a rough estimate of the refractory period.

The refractory period of the amoebae is about 2–4 min, depending on the stage of the aggregation process (this time decreases as aggregation proceeds), and this sets a lower limit to the rotation time of a rotor, which cannot be less than the refractory period. The refractory state of the amoebae is due to a desensitization of cAMP receptors by phosphorylation so that they become unresponsive to this signal until dephosphorylation occurs (see Section 3.2.3). Since the distance between the successive arms of a spiral is the distance travelled by the wave during one cycle of the central rotor, it is possible to calculate the rate of wave propagation from this distance and the rotor period.

▷ Given a rotor whose period is 5 min and a distance of 2 mm between the arms of a spiral in Figure 7.22, what is the velocity of propagation of the wave?

▶ 2 mm/5 min = 0.4 mm min$^{-1}$ = 6.6 μm s$^{-1}$

As aggregation proceeds, amoebae move towards the centre whose influence they fall under and the initially uniform lawn of amoebae is partitioned into a discrete sets of cells, as shown in Figure 7.23.

▷ What property of the cells is responsible for generating these discrete domains?

▶ Their refractory period after receiving a signal, which results in wave annihilation along the line where the waves meet.

Notice that the discrete domains forming in Figure 7.23 are separated by more or less straight lines, which are defined by lines that arise through the mid-point and at right angles to the lines joining adjacent centres. Given that the centres form in a fairly uniform distribution over the whole domain of cells, the result is a

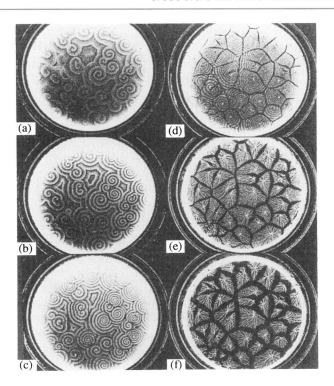

**Figure 7.23** Spiral aggregation patterns in the slime mould *Dictyostelium discoideum* gradually breaking up into separated streams, each of which produces a slug (see Figure 7.21).

5 cm

partitioning of the cells into roughly equal-sized aggregates of several thousand cells each. Each aggregate forms a mound of cells that piles up and then topples over, forming a multicellular 'slug' which migrates over the surface, leaving a slime trail behind it – hence its name, the cellular slime mould. As the slug moves, the cells differentiate according to their position: cells at the anterior end begin to change along a pathway that ends up as the stalk of the fruiting body (see Figure 7.21), while the majority of the cells, posterior to the tip, differentiate into the spore cells that end up as the ball of cells on top of the fruiting body. Periodic propagating waves continue to play a role in the dynamic organization of these processes of slug migration and cell differentiation. The waves are initiated at the tip of the slug and propagate towards the rear, cells moving forward within the slug with a periodic change of velocity as they respond to and transmit pulses of cAMP, the frequency being about 1 wave every 2 min.

Florian Siegert and Kees Weijer, in Munich, have shown that the generator of the periodic waves is, once again, a rotating activity wave at the tip. However, this is now initiating a wave in a three-dimensional structure, so the geometry of the wave is somewhat more complicated than the spirals in a plane that occur during aggregation when the cells are spread out on the agar surface. What Siegert and Weijer have deduced from the detailed observations of individual cell movements and the wave pattern is what is called a scroll wave, as shown in Figure 7.24. Looking at the tip, you would see a rotary wave propagating in the direction of the arrow on the circle, with cells moving in the opposite direction, towards the signal source. So cells rotate around the tip at right angles to the forward movements of the slug. But as the wave travels back along the slug, it changes its shape and transforms into a plane wave that propagates backwards through the slug, with cells moving forwards and giving coordinated movements to the whole.

This dynamic spatial pattern is connected in some way with the differentiation of anterior cells into stalk cells and posterior cells into spore cells, involving differential gene activation. It appears that frequency, phase, and amplitude modulation of the propagating wave are all involved in this regulation of cell

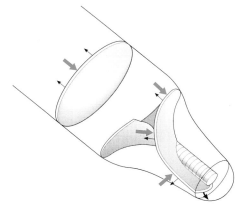

**Figure 7.24** The pattern of signalling and cell movement in the *Dictyostelium discoideum* slug during migration, showing a spiral wave at the tip changing into a plane wave that propagates backwards (small arrows) as the cells move forwards (large arrows).

differentiation but the details are still unclear. What is clear, however, is that the dynamic principles of physiological regulation that were encountered at the beginning of this chapter in connection with the growth and maturation of follicle cells in the ovary in response to periodic signals from the pituitary continue to hold for the 'simple' slime mould. But now space is added to time so that patterns of spatial order arise as part of the spectrum of emergent properties in excitable media.

The aggregation patterns of Figure 7.23 express one dynamic mode of an excitable medium made of signalling, moving cells. The wave pattern and movement of a slug represents a related but transformed pattern that is not unlike the coordinated beating of a heart with a pacemaker in a particular region, the wave of excitation and movement now producing movement of the slug across the surface instead of a contraction wave driving blood through the heart. The same principles of spatial and temporal organization apply to these two apparently very different systems, so that we can see how organs and organisms as excitable media express similar properties which emerge with different detailed characteristics in different contexts. And a dynamic mode that works well in one context, such as the multiple rotors generating the spirals of Figure 7.23 that lead to separate aggregation domains and then to slugs and fruiting bodies, is a disaster for the heart, which fibrillates to its, and the organism's, death in this mode. One of the great values of recognizing the operation of similar dynamic principles in excitable media, irrespective of their detailed properties, is precisely the identification of the limited set of dynamic modes available to them. These are the dynamic attractors of this class of dynamic system and they capture much of the behaviour of living systems.

### 7.3.6   An excitable medium that is not alive

Some of the patterns in Figure 7.22 are not spirals; they are concentric circles. So they cannot be generated by a rotating wave propagating in a closed loop, as in a spiral generator. How do they originate? The only possibility is that the cell at the centre is intrinsically rhythmic, so that it produces a pulse of cAMP every few minutes, initiating a wave. Notice that the concentric circles in the target patterns in Figure 7.22 are not so close together as the arms of a spiral: the rhythm of the pacemaker is slower than the cycling of a rotor. This is to be expected: the rotor is cycling close to the refractory period, which defines maximum frequency. The rhythmic pacemaker is pulsing more slowly. Such pacemakers are the initiators of aggregation, arising at random and releasing cAMP periodically. The presence of both types of wave tells us that slime mould amoebae have a dynamic mode available to them that most cardiac muscle cells do not, which is spontaneous rhythmic activity. However, because such pacemakers have a slower rhythm than the spirals, the latter take over as you can see in the adjacent spiral with a shorter wavelength.

This type of pattern of concentric circles was observed in a system with properties so unexpected that at first nobody believed the reports. In the 1950s a Russian biochemist, Boris Beloussov, mixed together some chemical reagents whose behaviour was intended to mimic the dynamic properties of the tricarboxylic acid cycle, the ATP-synthesizing cycle of cells described in Book 2, Chapter 4. To his astonishment and delight, the mixture changed its state periodically in time: Beloussov had discovered a chemical oscillator. Chemical reactions are not supposed to oscillate; they approach thermodynamic equilibrium monotonically. So Beloussov's papers, submitted to Soviet journals, were rejected, as they most likely would have been from European or American journals at that time. Chemists knew better than to believe such fairytales as oscillating reactions. However, sooner or later science catches up with its innovators and by the 1970s Beloussov's reaction, improved by his student A. M. Zhabotinsky, was recognized to have remarkable properties, and the Beloussov–Zhabotinsky (BZ) reaction became famous. Unfortunately, Beloussov died in 1970, on the threshold of this opening of chemistry and biology into the study of excitable media.

A Petri dish with a shallow layer of BZ solution is an example of an excitable medium that will spontaneously generate patterns in time and space, as shown in Figure 7.25. The original reaction mixture included inorganic and organic components such as bromous acid, sodium iodate, and iron(III) cyanide, together with malonic and ascorbic acids. What is happening is a periodic change of oxidation–reduction (redox) state which is driven by a process in which one molecule of bromous acid, HBrO, makes two molecules according to the following scheme:

$$HBrO_2 + BrO_3^- + 3H^+ + 2Ce^{3+} \rightarrow 2HBrO_2 + 2Ce^{4+} + H_2O$$

The cerium ion acts as an electron donor in this reaction. This reaction is inhibited by other products of the process, in particular by $CO_2$, so that it doesn't simply explode but increases and decreases rhythmically. Small volumes of the reagent do just this, the redox state changing periodically with a cycle time of a few minutes. This is made visible by a compound, ferroin, which changes colour from orange to blue as the redox potential changes. When the reagent is mixed in a shallow Petri dish, the periodicity is combined with wave propagation due to the autocatalytic reaction which spreads through the reagent, combined with diffusion of the products. The result is the target patterns seen in the photographs in Figure 7.25a, which were taken at 1 min intervals. The waves are propagating at about 1 mm min$^{-1}$, in the same range as the velocity of the waves in aggregating slime moulds. The patterns of concentric circles have the same dynamic explanation in both cases: the system is intrinsically rhythmic so that centres of activity can arise anywhere, periodically pulsing cells in the case of the slime mould and a periodic chemical reaction in the case of the BZ reagent. It is usually little heterogeneities in the dish, such as specks of dust, that provide surfaces where the centres form in the BZ reaction. It is evident that propagating waves annihilate one another when they collide, so the reaction has a refractory period: having swung to one end of the oscillation, the reagents have to be restored locally to particular concentrations before they can repeat the cycle.

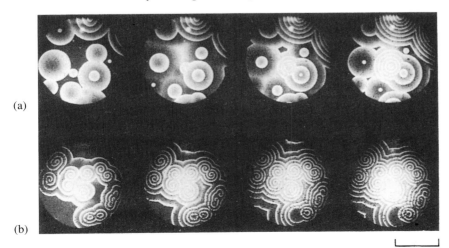

(a)

(b)

5 cm

*Figure 7.25* The same spatial patterns as those produced by aggregating slime mould amoebae, but in a dish of chemical reagent, producing the BZ reaction.

What about spirals? Figure 7.25b shows these. They can be initiated by gently stirring the solution, creating local heterogeneities and so generating conditions for rotors. These rotate with a period of about 2 min, again close to that of the slime mould spiral centres. So a remarkable similarity of dynamic behaviour is revealed between a chemical system and biological systems as diverse as aggregating slime moulds and beating or fibrillating hearts. What they share is the property of excitability, a dynamic condition that depends upon particular kinetic properties combined with spatial extension, and independent of molecular composition.

Such examples belie any notion that molecular composition is the fundamental level of causal explanation in complex systems. Rather, what needs always to be understood is dynamic order, which depends upon interactions between constituents of the system such as cells or molecules but is not explained by these. For example, simply knowing the chemical reactions that are involved, such as the production of bromous acid, is not enough to predict the behaviour of the system in time and space. It is necessary to study the total organization of the system as expressed in a mathematical model that puts the reactions and diffusion together into a coherent whole. Then the properties of the system emerge from this organization. To make this point even more clearly, Section 7.3.7 deals with emergent dynamic order in the behaviour of interacting organisms, which also show activity patterns characteristic of excitable media.

### 7.3.7   Emergent order in ant societies

Social insects, such as termites, bees, and ants, have remarkable levels of organization in their activities that arise from the cooperative, coherent behaviour of many interacting individuals. Scores of ants form living chains that pull together the edges of large leaves in trees of the tropical rainforests while others sew them together to make a nest. Termites construct their large, intricately sculpted colonial dwellings by turning the red dust of the savannah into rock-hard mounds with labyrinthine interiors, restoring damaged parts as if they had architectural plans of the whole. The dance of the bees is well-known, individuals sharing their knowledge of nectar sources with others by ritualized dance movements that indicate direction and distance to the flowers. The behaviour that will be considered in this section is perhaps not so dramatic as these examples, but it is typical of social insect organization and it reveals colonies as another case of excitable media with emergent order similar to those already examined, surprising as this may seem.

Ant species of the genus *Leptothorax* are very convenient for laboratory study. Colonies typically consist of 100 or so individuals and the ants are small so in their natural habitat they use a space such as the inside of a hollow acorn or narrow crevices in rocks within which to make their nests. In the laboratory they will happily make a nest in the space provided by two small squares of plastic, separated by thin strips of cardboard, with a source of sugar solution nearby. Under these conditions they can be easily examined. Blaine Cole, working in Houston, Texas, made videofilms of colonies consisting of different numbers of individuals and then analysed their movements. When there were only a few ants (up to 7) on the space provided, each individual had periods of activity and inactivity that had a chaotic pattern. The term 'chaos' is used here in its technical sense: individual ants have activity patterns that appear to be random but there is in fact some structure or order to the pattern that can be given a precise quantitative measure. The analysis of ant movements is very similar to that described in Figure 7.14, revealing a **chaotic** or **strange attractor** in the rhythmic pattern of a healthy heart. However, unlike the case of the heart, in ant behaviour there is no mixture of periodic and chaotic dynamics, when the ants are at low density (only a few ants in a defined space). Their activity–inactivity pattern is *solely* chaotic. What the ants characteristically do is move about for a period of time and then stop, remaining immobile for a while before moving again. Both their bouts of activity and their periods of rest appear to be of random duration, but actually fall on a chaotic attractor.

As the density of the colony was increased by adding more ants to a defined territory, Cole observed a sudden transition to dynamic order: patterns of activity and rest over the colony as a whole, measured as the sum of numbers of individual ants moving or resting, suddenly changed from chaotic to rhythmic with an average period of about 25 min. The difference between these activity patterns is shown in Figure 7.26. The graph in (b) shows the activity of a single ant, measured by the number of grid units in the territory over which the individual moves (measured by use of an automatic digitizing camera). The absence of any rhythmic components from this activity pattern is revealed in Figure 7.26d by a Fourier analysis of the data (a method of detecting periodicities in time series) from which no single frequency emerges. By contrast, in a colony of 10 or more ants, the pattern changes to that shown in Figure 7.26a, whose Fourier transform (Figure 7.26c) reveals a well-defined frequency peak at 25 min. The activity is still quite 'noisy', for some individual ants continue to be active when the majority are at rest, and conversely during activity bouts of the colony as a whole. Nevertheless, the transition from chaotic movement in isolated individuals or low-density colonies to rhythmic behaviour in a colony when the density is sufficiently high is perfectly clear. The group behaves in a collective mode that could not be predicted from the behaviour of individuals. This is a clear example of emergent behaviour. How are we to understand it?

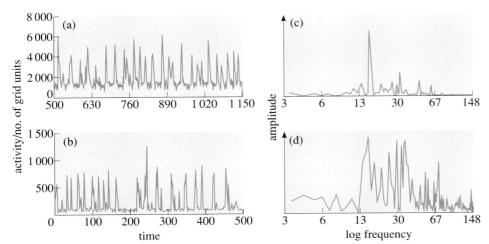

**Figure 7.26** Activity patterns in an ant colony (a), and in an individual ant (b). The former has rhythmic activity–inactivity cycles, as revealed by the Fourier transform with a peak at about 25 min (c), while the latter does not (d).

In biology, 'understanding' often means suggesting in what way some property contributes to survival, so that it can be understood as a useful property that has emerged through natural selection. Coordinated rhythms of activity in a colony could serve many functions: active teams engaged in collective work are more effective than chaotic individuals; and there are spatial patterns associated with the temporal rhythms, as is to be expected in excitable media, that could serve to organize activities in the colony. However, this type of functional explanation puts the cart before the horse. It implies that the phenomena we see in biology are there because they are useful. But there are many properties of organisms that are not of any obvious use, such as the structure of the retina in our eyes which requires photons to pass through nerve cells before being absorbed by molecules in the pigmented layer which then excite the neurons. Squids do it the other way round, which seems the sensible arrangement. So why do our eyes (and those of all vertebrates) have this curious structure? Because the sequence of events that gives rise to eyes during embryonic development in vertebrates involves foldings and bucklings of sheets of cells such that the epidermis that produces the pigmented cells ends up on the inside of the retina; and this works as a visual system. The message here is that explanations come from understanding dynamics – how the system is generated – and then understanding function within this context. Function is always secondary to structure: something has to be generated before it can be used. So now it is necessary to try to understand how rhythmic activity in ant colonies emerges from the chaotic behaviour of individuals. This requires a model.

### 7.3.8   Modelling an ant colony

Why should density of ants play an apparently crucial role in the transition from chaotic to ordered behaviour? Ants interact with one another, and an active ant encountering an inactive one will stimulate the latter into movement. At low densities there are few encounters, but at higher densities activity can spread like a contagion through a colony. This seems to be the simple key to the process and it is just like the spread of activity in an excitable medium. Now ants become the excitable elements.

Using this approach, Octavio Miramontes from Mexico, working with Brian Goodwin at the Open University in collaboration with Ricard Solé from Spain, constructed a model of ant colony behaviour that provided useful insights into the dynamics of this system. Individual ants were modelled as elements driven by chaotic dynamics. When in an active state, individuals moved from one site to any adjacent, unoccupied site on a lattice that defines the territory of the colony (Figure 7.27). If an individual moves to a site adjacent to one occupied by an inactive ant, the latter is stimulated to become active, but its activity will cease after a time unless it becomes spontaneously active or is stimulated by another ant. Because the elements ('ants') obey simple logical rules, they belong to a category of models which are known as cellular automata and in this case they are mobile cellular automata. These are very useful in studying the properties of complex dynamic systems of various kinds, made up of interacting elements.

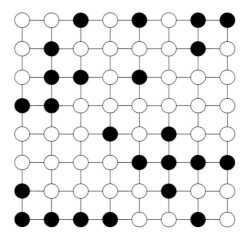

**Figure 7.27**   A grid describing a colony of model ants, black circles representing occupied sites and white circles unoccupied sites of the grid.

In the model, 'colonies' of one or a few individuals had chaotic movement patterns, as shown in Figure 7.28a, with the corresponding Fourier spectrum showing a broad range of frequencies (Figure 7.29a). However, above a critical density there was a transition to rhythmic activity over the 'colony' (Figure 7.28b), which became more clearly defined as the density increased (Figures 7.28c and d). Figures 7.29b, c and d are the corresponding Fourier spectra, showing the emergence of a dominant frequency throughout the colony. At higher densities the rhythm becomes very pronounced and regular, as in (e) in both figures, but this is beyond the range of densities normally encountered in real colonies.

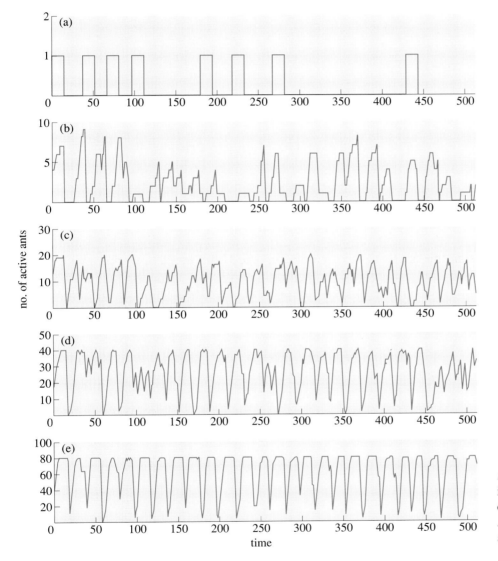

**Figure 7.28** Patterns of activity–inactivity in model colonies with different numbers of ants in the same territory. As the density increases, a rhythm appears, just as in real colonies.

What is the nature of this rather sudden change in the pattern of colony activity, from disorder to order? This type of phenomenon is well known in physical systems, as in the condensation of a gas to a liquid just below the boiling point or the sudden emergence of a magnetic field in a ferromagnet when it cools below a critical temperature. However, the ant colony is 'heating up' rather than cooling down, in the sense that as the density increases the amount of activity per individual increases, due to increased interactions. To get some insight into the nature of the transition, we need a measure that tells us something about order–disorder relations. An obvious quantity to use is entropy.

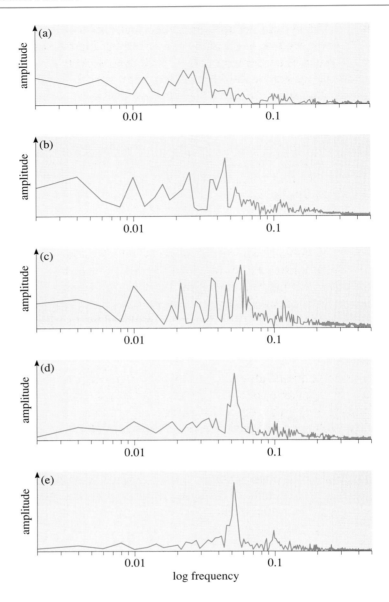

**Figure 7.29** The Fourier transforms of the activity–inactivity patterns of Figure 7.28, showing the emergence of a clear peak corresponding to a well-defined rhythm in the model as density increases from (a) to (e).

▷ Recall from Book 2, Chapter 2, the definition of entropy and its relationship to order in physical systems.

▶ Entropy can be interpreted as a measure of the disorder in a system, or the tendency for all possible dynamic modes to be equally occupied.

An appropriate measure of entropy for a dynamical system such as an ant colony is

$$S = -\sum_{i=0}^{N} p_i \log p_i$$

where $N$ is the total number of individuals in the colony and $p_i$ is the probability of finding $i$ out of the $N$ individuals in an active state. When this quantity is determined for colonies of different density, the result is shown in Figure 7.30. The entropy goes through a peak at a density of 0.24, which is where the transition occurs (starting at 0.2). As the density is increased from very low values the entropy increases, which means that the diversity in the number of active ants increases; that is, the system occupies more and more of the dynamic modes available to it, until they are all occupied. The diversity increases to the point where activity can percolate throughout the whole colony. At this point colony-wide order suddenly appears in the form of a collective rhythm, reflecting long-range coherence. As density is increased further the entropy decreases again, order becoming established in the form of regular oscillations while the diversity in the number of active individuals decreases, since the colony approaches the condition where all its members are either active or inactive at any one time – a condition of high dynamic order and low entropy. So this analysis gives some insight into the reasons for the sudden emergence of rhythmic activity, which in the model is a real phase transition.

Can something be said about where actual ant colonies are located on the scale of densities? Very interesting studies of *Leptothorax* colonies by Nigel Franks at the University of Bath, who was the first to report rhythmic behaviour patterns in ants, suggest that colonies construct nests of a size that allocates a nearly constant area of about 5 mm² per individual so that colony size is proportional to numbers of ants in the colony. As ants come and go from the nest, changing the density, rhythmic activity patterns appear and disappear, replaced by disordered activity. It appears that the colonies are regulating their densities so that they live on the edge of chaos.

This rather dramatic phrase, 'life at the edge of chaos', has entered the description of biological systems in terms of complexity theory, the use of non-linear models involving chaos and emergent order to describe biological processes of the type considered in this chapter. The basic idea is that in a continually changing world the best dynamic strategy to adopt is one that keeps all the dynamic modes accessible so that they can be continually assessed for their relevance to changing circumstances. This is the edge of chaos, where, as we have seen, the entropy – the dynamic diversity available to the system – is at a maximum. Just in this region, order emerges from chaos and organisms make use of this order, as in the cooperative activities of social insects.

Not only does temporal order emerge in the form of a colony-wide rhythm, but also spatial order, as is to be expected in an excitable medium. Figure 7.31 shows the distribution of activity levels in the colony territory of the model, showing a pattern in which activity levels increase towards the centre. Ant colonies are spatially organized with the brood (developing larvae) of different ages distributed in concentric circles with the queen (if there is one) at the centre and the oldest larvae towards the edges. The model certainly does not explain this distribution, which involves larval age rather than activity; but it does show that a sense of spatial location can accompany the emergence of order, and could be used to organize spatial aspects of the colony.

As conditions change and the colony grows or shrinks depending upon circumstances, the organization of space and activities need to change in appropriate ways. Living on the edge of chaos means being just a little way into the ordered regime so that, as the density changes in the colony due to fluctuations in population numbers (ants coming and going, being born and dying), patterns of order dissolve into chaotic dynamics and then get re-established in a manner that is responsive to current circumstance. The edge of chaos may be a generic attractor for a whole class of complex dynamic systems that includes living systems as their most familiar representatives. From these insights a theory of life processes that is fully integrated with physics and chemistry, to which this course is dedicated, could emerge; and it, too, will live on the edge of chaos.

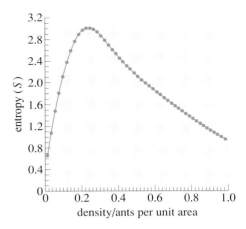

**Figure 7.30**  Change in the entropy of the model colony as a function of density, showing the occurrence of a maximum at the point where a chaotic pattern undergoes a transition to a rhythmic pattern.

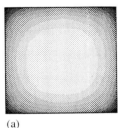

(a)

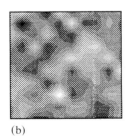

(b)

**Figure 7.31**  (a) The spatial pattern of mean activity in a model colony, showing a concentric distribution with maximum activity toward the centre. (b) The colony without interactions between ants, showing an absence of spatial pattern.

## Summary of Sections 7.3.5–7.3.8

Cellular slime moulds have a life cycle that involves a change of state from single independent cells to a multicellular organism. During the transition, cells communicate with one another by periodically releasing and responding to pulses of cAMP, which induces cell movement. A population of these rhythmically communicating cells has the properties of an excitable medium, and waves of cell movement with both concentric circle (target) and spiral patterns are observed. In the multicellular form (the slug) the periodic waves of activity continue and result in pulsatile movement of the slug. The rhythmic cAMP pulses involved are somehow connected with the differentiation of the slug into the two cell types that make up the fruiting body.

A purely chemical system, called the Beloussov–Zhabotinsky (BZ) reaction, can also spontaneously generate patterns of concentric circles and spirals that propagate through the solution. These are due to periodic changes in the concentrations of the reacting chemicals.

Ant colonies, too, are self-organizing, generating order in space and time. In particular, rhythmic activity patterns have been observed in colonies when the density of ants is above a particular value, while at lower densities the activity is chaotic. A simple model of a colony as an excitable medium in which ants are described as mobile elements that obey simple rules of interaction reveals a similar transition from chaos to order at a critical density. This is technically a phase transition and the entropy of the 'colony' has a maximum at this point, which has come to be called 'the edge of chaos'. A conjecture has emerged that this is the state to which complex dynamic systems will naturally tend, as suggested by the maximum entropy condition, and that it results in flexible adaptive behaviour, relevant order emerging out of chaos.

## Objectives for Chapter 7

After reading this chapter, you should be able to:

**7.1** Define and use, or recognize definitions and applications of, each of the terms printed in **bold** in the text.

**7.2** Give examples of the range of frequencies encountered in the rhythmic activities of organisms.

**7.3** Present evidence for oscillatory behaviour and frequency control as a basic mode of communication and regulation in the menstrual cycle.

**7.4** Describe simple harmonic motion in an ideal pendulum in terms of energy conservation, and be familiar with the terms used to describe oscillatory activity.

**7.5** Explain the difference between an ideal and a real pendulum in terms of energy, and understand the concept of an attractor.

**7.6** Describe the factors involved in the synchronization and entrainment of oscillators, and use these to analyse the interactions between organisms and with their environments.

**7.7** Define the properties of biological clocks and circadian rhythms.

**7.8** Show how to analyse the dynamic properties of the eclosion clock in *Drosophila*, explaining how and why it can be stopped.

**7.9** Describe the results of genetic analysis in the study of biological clocks and explain how the design of genetic experiments influences the results obtained.

**7.10**   Explain how ventricular fibrillation and sudden cardiac arrest may be explained in terms of the heart as a dynamic system.

**7.11**   Define an excitable medium and describe examples in chemistry and biology that illustrate its characteristic dynamic propeties.

**7.12**   Explain how the notions of excitability and communication can be applied to ant colonies to explain the emergence of rhythmic activity patterns.

## Questions for Chapter 7

### Question 7.1   *(Objectives 7.2 and 7.3)*

Which of the following statements are true?

(a)   Rhythms with different frequencies in organisms are always coupled together so that they have well-defined frequency ratios.

(b)   In the absence of periodic signals from the external environment, groups of organisms can become synchronized to one another's rhythmic activity and so develop a group rhythm.

(c)   The primary source of the menstrual rhythm is from positive and negative feedback signals arising from the growing follicle and the corpus luteum.

(d)   In the menstrual cycle, all physiological rhythms involved in controlling the process have periodicities that are about 28 days or small fractions thereof.

(e)   Since many physiological processes are rhythmic, it would make sense to administer drugs periodically in synchrony with the rhythms so as to produce the maximum dose.

### Question 7.2   *(Objectives 7.4 and 7.5)*

Draw a trajectory that describes qualitatively the motion of a real clock pendulum that is started from rest with a small impulse so that the initial amplitude of the pendulum is *smaller* than the final amplitude.

### Question 7.3   *(Objectives 7.6 and 7.7)*

You are given a culture of *Euglena*, a flagellated unicellular photo-autotroph that lives in fresh water. You want to investigate the origins of a daily rhythm in cell movement in which cells occupy the upper 2–3 cm of the culture during the day and descend to a depth of 8–10 cm during the dark hours. How would you design experiments to:

(a)   test for the presence of an internal circadian clock;

(b)   see whether the cells influence one another in maintaining a regular rhythm;

(c)   find out if the circadian rhythm is temperature-compensated?

### Question 7.4   *(Objectives 7.8 and 7.9)*

A new mutation is discovered in *Drosophila* that alters the period of the eclosion rhythm from its normal period of about 24 h at 20 °C to 40 h. The mutant can still be phase-shifted by light signals. It is observed also that the developmental rate of the mutant larvae is decreased so that the larval period is increased from the normal 24 h to 38 h, while the pupation period is also extended, from 4 days to 6 days. Previous evidence indicated that the circadian clock controlling eclosion has no effect on the pupation process itself, since the clock can be stopped by a light flash at a critical phase in the clock cycle (see Section 7.2.3), without affecting the development of the pupa into an adult, its only influence being to determine the time of day when the adult emerges from the pupal case. How would you explain these apparently contradictory observations?

*Question 7.5* *(Objective 7.11)*

**Figure 7.32** For use with Question 7.5.

What's wrong with Figure 7.32? Explain.

*Question 7.6* *(Objective 7.12)*

Ants observed outside their nest, engaged on various tasks such as foraging and patrolling, do not show any obvious overall group activity rhythm of the type observed in the nest. However, there is some evidence of rhythmic activity waves that propagate along columns of ants in foraging trails. How would you explain such observations?

# Answers to questions

## Question 1.1

For the *E. coli* bacterium the gut is a foreign external environment, liable to fluctuations in conditions over which it has no control, but which it may be required to adapt to if it is to survive. For the liver cell the environment is the circulating extracellular fluid derived from the bloodstream, which is rather precisely buffered and its composition controlled by homeostatic mechanisms present at the level of the organism as a whole; it therefore does not need to possess the whole adaptive panoply of mechanisms that the *E. coli* bacterium requires.

## Question 1.2

Such a sensitive thermostat would respond very rapidly to minor fluctuations in room temperature and therefore would be in continuous oscillation, switching the heating system on and off many times a minute, leading to inefficiencies and possible breakdown.

## Question 1.3

(a)  Whichever is the slowest step in the pathway at any particular time; this is often the first in the sequence, or the first beyond a branch point in the sequence. In the branched pathway shown it would be more functional to regulate the two branches separately, so one might expect the main control points to lie between A and B and also between C and D and C and G.

(b)  End-products F and K might inhibit enzyme A-ase and F might inhibit the C → D reaction whilst K would be expected to inhibit the C → G reaction (Figure 1.12). This inhibition beyond the branch point is the more sensitive.

(c)  A common mechanism would be to phosphorylate or dephosphorylate serine or threonine residues on the enzyme protein by means of an A-ase kinase or an A-ase phosphatase; such reversible covalent modification will affect the kinetics of the A-ase reaction.

The A-ase kinase and phosphatase may themselves be activated (or inhibited) by kinases or phosphatases earlier in the signalling cascade (recall Figure 1.6) or by signalling molecules or ions such as cAMP or $Ca^{2+}$. (Chapter 3 deals with such signalling pathways in more detail.)

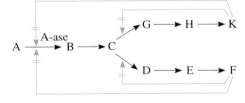

**Figure 1.12**   Answer to Question 1.3b.

## Question 1.4

The operon. This mechanism of control is very important in unicells, with their fluctuating external environment and food sources. It is much less significant in the constituent cells of multicellular organisms whose environment is much more constant.

## Question 2.1

Equations 2.3 and 2.4 predict that in a time $t$ the ion will move a distance $r_1$ due to viscous flow under the influence of the electric field where

$$r_1 = \left(\frac{D}{kT}\right)eEt$$

In the same time the average distance it will diffuse is given by Equation 2.1 as $r_2$ when

$$r_2 = (6\,Dt)$$

For short times the distance $r_2$ is greater than $r_1$ and for long times the reverse is true. Thus the required minimum time is obtained by equating $r_1$ and $r_2$ to obtain

$$t = \left(\frac{6}{D}\right)\left(\frac{kT}{eE}\right)^2$$
$$= 282\text{ s}$$

### Question 2.2

At 25 °C, $kT = 4.11 \times 10^{-21}$ J. The electrical energy is

$$qV = (1.6 \times 10^{-19}) \times (0.1) = 1.6 \times 10^{-20}\text{ J}$$

Thus the ratio of the electrical energy required to the typical thermal energy is

$$\frac{1.6 \times 10^{-20}}{4.11 \times 10^{-21}} = 3.89$$

### Question 2.3

Using Equation 2.9 and the data from Table 2.2, the inside of the cell will, at equilibrium, reach a voltage of +60 mV, −102 mV and −101 mV when the membrane is permeable to $Na^+$, $K^+$ and $Cl^-$ ions respectively. A leak across the membrane of $K^+$ or $Cl^-$ ions would have little effect on the resting membrane voltage but a leak of $Na^+$ ions would drive the resting membrane voltage positive from −100 mV to +60 mV.

### Question 2.4

The membrane voltage $V_m = -0.060$ V and the Nernst potential (Equation 2.9) is −0.052 V. Thus, from Equation 2.13, the change in the electrochemical potential of the $K^+$ ion on entering the cell, $dg_{in}$, is negative so that there will be a spontaneous flow of $K^+$ ions into the cell when the channel opens.

### Question 2.5

In order to function spontaneously the sum of the changes in electrochemical potential of the three ions must *decrease* when the coupled transfer is made. Bearing in mind that for the driven ions $dg_{out} = -dg_{in}$, this requires that

$$z_1(V_m - V_{N1}) - 2z_2(V_m - V_{N2}) < 0$$

### Question 2.6

See Figure 2.16. When $P_2 = 0.5$, $P_{on} = (0.5)^4 = 1/16$ so that one channel in 16 is open at the membrane voltage at which one half of the equilibrium gating charge has been transferred. Increasing $fq$ increases the maximum slope of the curve.

### Question 3.1

Both are modes of passive transport, which rely on the presence of a concentration gradient across the cell membrane; the greater the concentration gradient, the faster the rate of movement down the gradient. (If a substance is charged, then the

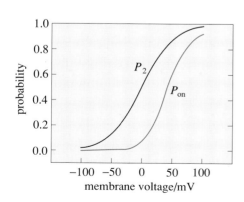

**Figure 2.16** Graphs of $P_2$ and of $P_{on}$ for an assembly of identical channels, as a function of $V_m$.

rate and direction of movement also depends on the electrical gradient across the cell membrane.) The two processes can be distinguished by kinetic studies: facilitated transport exhibits saturation, whereas the rate of transport by simple diffusion is always proportional to the concentration gradient across the cell membrane.

Transport may be enhanced by the presence of intracellular binding proteins which complex with the transported substance. This maintains a concentration gradient by lowering its free intracellular concentration. Substances that are hydrophilic are unable to simply diffuse across a lipid bilayer since they do not interact favourably with its lipid core, so these compounds rely on specific proteins in the cell membrane such as passive facilitated transporters in order to permeate the membrane. In contrast, hydrophobic substances are readily diffusible in phospholipid and so may diffuse across without the requirement for a specific transport protein.

## Question 3.2

(b) and (c).

## Question 3.3

The hydrophobic structure enables the transporter to favourably associate with the lipid bilayer as an integral membrane protein. The hydrophilic regions are important for interacting with the aqueous extracellular and intracellular media (including the solute transported).

## Question 3.4

The aspartate which had been removed from the protein played a crucial role in the ATPase transporter mechanism. Usually, the aspartate is phosphorylated (by ATP) when $Na^+$ is present, and this process initiates the conformational changes which result in $Na^+$ (outward) and $K^+$ (inward) movement across the cell membrane.

## Question 3.5

(a) False: receptors that are also ion channels or enzymes do not require intermediary G proteins, since they are regulated directly via ligand binding.

(b) True: The ion channel both receives and carries out the response to the stimulus (transmitter).

(c) True.

(d) False: the nicotinic receptor is a transmitter-gated ion channel which is immediately activated by ligand binding; in contrast, the muscarinic receptor produces a slightly slower response as it operates via an intermediary G protein which then in turn binds to an amplifier.

(e) False: hydrophobic ligands (e.g. steroids) can cross the cell membrane and bind to intracellular receptors.

(f) False: ligand–receptor binding involves non-covalent, reversible interactions such as hydrogen bonds and electrostatic attraction.

## Question 3.6

(a) At a neuromuscular junction; (b) in photosensitive cells in the eye; (c) in electrically active tissues such as nerve cells.

## Question 3.7

(a) See Figure 3.64.

(b) Many molecules of G protein can be stimulated in quick succession by an activated receptor and many molecules of second messenger can be produced before auto-inactivation.

(c) The concentration of agonist decreases, so that the G protein is no longer stimulated. Those G protein subunits already activated revert back to their inactive form following GTP hydrolysis by the α subunit's intrinsic GTPase. As a result, the amplifier is no longer stimulated.

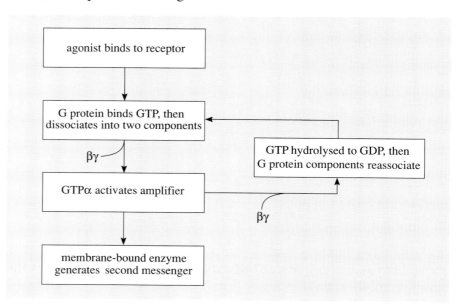

**Figure 3.64**  Answer to Question 3.7a (completed Figure 3.63).

## Question 3.8

(a)  Incubate target cells with a non-hydrolysable GTP analogue such as p(NH)ppG. If a G protein is involved, then the biological response of hormone Z should be observed (possibly with greater magnitude).

(b)  The addition of bacterial toxins could provide further information on the type of G protein involved. If the biological response is reproduced in the presence of cholera toxin only, then this suggests the involvement of $G_s$; whereas if pertussis toxin blocks the biological response of hormone Z, then $G_i$ is likely to mediate the intracellular signalling.

(c)  The G protein would be artificially activated, causing a biological response.

(d)  $G_s$, adenylyl cyclase and cAMP (Figure 3.46).

## Question 3.9

(a)  (i) and (ii) adenylyl cyclase; (iii) cGMP phosphodiesterase; (iv) phospholipase C

(b)  Tyrosine kinase and guanylyl cyclase are activated by direct ligand binding as these are found on the cytosolic domains of receptors (Figure 3.25). In addition, soluble (cytosolic) guanylyl cyclase is activated by nitric oxide.

## Question 3.10

The following could be associated with an activation of phospholipase C: (a), (b), (d), (f), (g), (h), (i), (j) and (l). See Figure 3.65.

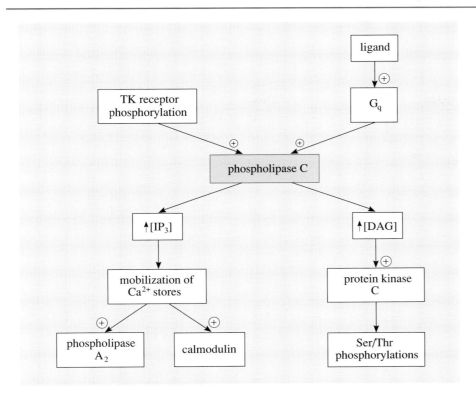

***Figure 3.65*** Answer to Question 3.10.

## Question 3.11

(a) Acetylcholine is more potent: as $pD_2$ is the negative logarithm of the concentration required to produce a half-maximal response, the greater the value, the more potent the agonist.

(b) Hexanoyl choline is an antagonist which blocks the actions of acetylcholine at its receptor, thereby lowering the $pD_2$ value.

## Question 4.1

The synapse is the site where chemical signalling occurs in the nervous system. Action potentials travel down the axon to the synapse, where they trigger the movement of synaptic vesicles packed with neurotransmitter to the synaptic junction where they fuse with the membrane releasing the transmitter into the synaptic cleft. The transmitter diffuses across the synaptic cleft and binds to one of many specific receptors located on the post-synaptic membrane, opening channels for the entry of calcium ions, which depolarize it.

## Question 4.2

Vesicles containing neurotransmitter are found at the pre-synaptic nerve ending only and the receptors for them are located exclusively at the post-synaptic membrane. Hence, sending an electrical impulse in the opposite direction would not enable the signal to pass across the synapse.

## Question 4.3

The release of TRF from the hypothalamus and TSH from the pituitary gland is under negative feedback control by the plasma thyroid hormone concentration. In an iodine-deficient individual, thyroid hormone concentrations will be low, so in response to this TRF and TSH levels would be increased. However, they will be unable to stimulate the thyroid gland to produce hormone because of a deficiency in iodine.

## Question 4.4

Since urine flow increases following alcohol consumption, vasopressin must have a decreased reabsorptive effect on the kidney. This could be due to alcohol causing an inhibition in the secretion of vasopressin from the pituitary gland, or an increased rate of removal of vasopressin from the bloodstream or a decreased sensitivity of target cells in the kidney to the hormone.

In fact, the major effect of alcohol on urine flow is due to its ability to inhibit the secretion of vasopressin.

## Question 4.5

One example described in this chapter is the interaction between a B cell and a helper T cell (see Figure 4.12), in which the B cell acts as an antigen-presenting cell (APC). The B cell engulfs the intact antigen and processes it into peptide fragments, which associate with MHC II molecules within the cell. This assembly is externalized, presenting the helper T cell with a complex ligand on the surface of the B cell, consisting partly of sequences in a peptide fragment of the antigen and partly of sequences in the MHC molecule. The T cell receptor binds with both components of this ligand. The two cells are also held in close contact by interactions between adhesion molecules and their receptors on opposing cell surfaces. While in such close contact, short-range chemical signals pass between the two cells, which increase the density and affinity of several receptor–ligand interactions between opposing cell membranes. When a 'communication threshold' is reached, the helper T cell transmits a cytokine (in this case one of the interleukins), which gives the final activating stimulus to the antigen-primed B cell to differentiate into an antibody-secreting plasma cell (see Figure 4.13).

## Question 4.6

It is synthesized from a readily available precursor and can easily diffuse away from its site of production due to its small size and hydrophobic nature. However, its unstable property ensures that it acts only in the immediate vicinity.

## Question 4.7

(a), (b) and (d) are correct. (c) is false since there appears to be no selectivity for charged substances that can move between cells by this route. The main discriminatory factor is size: substances larger than $M_r$ 1 000 cannot move into adjacent cells, probably because of the physical constraints of the pore.

## Question 4.8

(c) Arginine.

## Question 4.9

(a) Both forms of chemical signalling involve receptor molecules which are located in target cells.

(b) Both encompass a broad spectrum of chemical compounds, which have either hydrophilic or hydrophobic properties.

(c) Hormones have a longer half-life than autocrine and paracrine factors.

(d) Hormones are synthesized and stored in specialist glands and are released into the bloodstream in response to appropriate stimuli. Local mediators are synthesized in a diverse range of cell types. They are immediately produced (without storage) following stimulation, and move through the extracellular space, acting on cells in the vicinity without a reliance on the circulatory system.

## Question 4.10

Both drugs inhibit eicosanoid synthesis in tissues. Steroids such as cortisol induce the formation of lipocortins inside cells and these compounds in turn decrease the activity of phospholipase $A_2$, the enzyme that generates the arachidonic acid precursor. Drugs such as aspirin and indomethacin act further downstream in the eicosanoid pathway and inhibit cyclooxygenase.

## Question 5.1

In the case of the chain of cells there is a high resistance $R$ in the signal path caused by the narrow junctions connecting the cells. This reduces the current that can flow between connected cells. As a result, the time constant, $t_{RC}$, for charging the cell (which by definition equals $RC$ and so is proportional to the resistance $R$) is increased; hence the propagation speed is low. In the case of the axon, the whole axoplasm, which has a much lower resistance than the gap junctions, connects neighbouring sections of the axon and so propagation speeds are much higher.

## Question 5.2

The speed of propagation of an action potential along a chain of cells is proportional to the cell diameter, $d$, and inversely proportional to the time, $t_{RC}$, taken to charge the cell. As $t_{RC} = RC$ by definition, the speed will be proportional to $d/RC$. But as $R$ is assumed to be constant and $C$ varies as $d^2$, the speed will vary as $1/d$. (This assumes that the voltage rise of the cell is fast when the ion channels in the cell membrane open.)

## Question 5.3

(a) Blocking some $Na^+$ channels would reduce the magnitude of the initial positive voltage rise in the axon and would also reduce the rate of rise. If the voltage rise were not sufficient to open neighbouring $Na^+$ channels, the action potential would stop rapidly.

(b) Blocking some $K^+$ channels might increase the magnitude of the initial positive voltage rise because of the reduced flow of positive $K^+$ ions outward across the membrane. The recovery period before the axon could propagate another signal would also be prolonged, even to the point where recovery might be halted. Hence the minimum time between the propagation of successive action potentials would increase, possibly so much so that action potentials would effectively cease altogether.

## Question 5.4

The time taken to diffuse a given distance, $d$, is proportional to $d^2$, therefore diffusion is fast over short distances but becomes very slow over longer distances. Ionic diffusion also spreads uniformly outward from a single source, and so reaches all of the interior of the cell in due course; this may be satisfactory for signals within one cell, but communication between cells often needs to be more specific and capable of rapidly targeting individual cells. For this purpose either a chain of connected cells or a connecting nerve cell is required.

### Question 6.1

If the slices are to be distinguished, the magnetic field at the centres of adjacent slices must differ by at least $2 \times 10^{-4}$ T. But the slices are separated by 1 mm in the $x$-direction so that the magnetic field gradient must be at least $2 \times 10^{-4}$ T/ $10^{-3}$ m $= 0.2$ T m$^{-1}$.

### Question 6.2

Living systems may have evolved to use or to exclude the types of fields that occur naturally, but they have not had the time to react to the human-made electromagnetic environment. For this reason, radio-frequency fields may be a more likely source of cell damage than static fields.

### Question 7.1

(a) False. Heart-beat and sleep–wake cycles, for example, are not synchronized with one another though the average heart *rate* does vary between day and night.

(b) True. See Section 7.1.2.

(c) False. The positive and negative effects of oestradiol and progesterone, respectively, on the pituitary are only part of the control of regulatory activities involved in menstrual cycle periodicity.

(d) False. The rhythm of gonadotropin release from the pituitary (about 1 cycle h$^{-1}$) does not have a simple frequency relationship to the menstrual period.

(e) True. This is now becoming a widely used method of drug administration. Smaller amounts of a drug are required when it is given during the rising phase of the endogenous rhythm of the process it is intended to affect.

### Question 7.2

The initial impulse starts the pendulum moving from the point marked S (start) in Figure 7.33. As the pendulum swings through an initially small amplitude orbit (marked 1), it loses energy so that at the end of the first cycle it ends up closer to the origin (O) than S. The escapement mechanism then gives the pendulum an impulse that takes it out to the orbit marked 2. At the end of this cycle the pendulum has an amplitude greater than the initial point, S, and it then gets another impulse that takes it to orbit 3. This process of increasing amplitude continues until the pendulum reaches an orbit where the loss of energy during a cycle just balances the energy coming from the escapement mechanism, and the clock reaches its stable limit cycle.

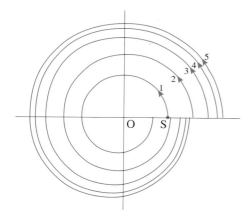

**Figure 7.33** The motion of a real pendulum that is started at S with an amplitude smaller than its final amplitude.

### Question 7.3

You need first a method of measuring cell density at different depths in the culture vessel which contains water to a depth greater than 10 cm. The culture can be sampled with a pipette at depths of, say, 2 cm and 8 cm, the density of cells in the sample being measured by either optical density or by a cell-counting method. Measurements should be taken frequently, at least once per hour around the clock, and from different positions in the culture vessel at the two designated depths. Each reading is then the average of the values at the different positions and constant depth.

(a) To test for an internal circadian clock, all controllable environmental variables, particularly light and temperature, should be kept constant. Light must be maintained at some low intensity that allows the cells to photosynthesize. Then the culture is sampled to see if there is a persistent rhythm of movement. If there is, it will be at a periodicity that is different from, but close to, 24 h.

(b) If the rhythm of cell movement under constant light conditions persists indefinitely with a significant amplitude, then it is very likely that cells are influencing one another so that they retain a degree of synchrony. Otherwise the collective rhythm would gradually disappear and each cell would continue to rise and fall with its own rhythm, variations between cells resulting in a complete loss of a collective rhythm after many days under constant conditions.

(c) To investigate temperature compensation, cultures are transferred from LD12 : 12 at a standard temperature (e.g. 21 °C) to constant dim light at a different temperature, say 25 °C. If the circadian period at 25 °C is not significantly different from that at 21 °C, or if it is longer, then the clock is temperature-compensated.

## Question 7.4

Winfree's demonstration that the circadian clock can be stopped by a critical light stimulus delivered to genetically normal pupae depends upon a purely dynamic propety of the clock – the existence of the point P in Figure 7.10 where the clock has no time and no motion. All the molecular components of the clock in these pupae are intact and normal, so there is no reason for there to be any effect of such an experiment on any other aspect of the pupae – in particular, their development to the stage where they are ready to emerge as adult flies from the pupal cases.

However, a mutation that affects the period of the circadian clock significantly, such as the (fictitious) one described that results in a 40 h free-running period, involves a change in some molecule that is basic to clock function. This is likely to affect metabolism and development as well, such as slowing down the rate at which the organism develops so that pupation takes longer, as described. A mutation of this is like the chol-1 mutant in *Neurospora*, which simultaneously affects growth rate and circadian period.

## Question 7.5

Spiral waves can arise in excitable media but water is not excitable and so the spiral wave cannot occur. Try it.

## Question 7.6

The density of ants outside the nest is, on average, much lower than it is in the nest, so the critical density is below that required for a colony-wide activity rhythm to occur. However, ants in foraging trails are at densities at which rhythms could occur, and these would be expected to propagate spatially as in any excitable medium.

# *Further reading*

*Chapter 3*

Berridge, M. J. (1993) Inositol trisphosphate and calcium signalling, *Nature*, **361**, pp. 315–325.

> A comprehensive review article which consolidates many aspects of calcium signalling described in this chapter and also provides interesting extension material.

Linder, M. E. and Gilman, A. G. (1992) G proteins, *Scientific American*, July, pp. 36–43.

> A well-illustrated and readable review.

Mathews, C. K. and van Holde, K. E. (1995) *Biochemistry*, 2nd edn, Benjamin/Cummings.

> This general textbook includes a number of topics covered in Chapter 3. The section on transport across membranes in Chapter 9 is superbly illustrated and supports the information given in Section 3.2 of this book.

*Trends in Biochemical Sciences* (1992), **17** (10).

> The entire October issue is devoted to the subject of intracellular signalling and contains a variety of detailed articles which provide more in-depth information about many of the topics covered in Chapter 3, including receptor ion channels, G proteins, inositol lipids, calcium and tyrosine kinase receptors.

*Chapter 4*

Ashcroft, F. M. and Ashcroft, S. J. H. (eds) (1992) *Insulin: Molecular Biology to Pathology*, IRL Press.

> An accessible introduction to the current research into all aspects of insulin biochemistry. It is written with the non-specialist in mind and includes an extensive glossary of useful terms. Section V provides extensive information on diabetes.

B-cell Signalling Special (1994) *Immunology Today*, **15**, No. 9, 393–454, Elsevier Trends Journals.

> A highly specialized compilation of recent conference reports and review articles devoted to B-cell signalling and its regulation; includes a full-colour 'pull-out' map of communication pathways involving B cells.

Knowles, R. G. and Moncada, S. (1992) Nitric oxide as a signal in blood vessels, *Trends in Biochemical Sciences*, **17**, pp. 399–402.

> A brief and highly readable account of nitric oxide, focusing on its role in smooth muscle relaxation.

Roitt, I. (1994) *Essential Immunology*, 8th edn, Blackwell Scientific Publications.

> An excellent general textbook of immunology, entertainingly written and well illustrated; Chapter 3 'Antibodies' and Chapter 4 'Membrane receptors for antigen' are also relevant to S327, Book 1, Chapter 3; Chapter 9 'Lymphocyte activation' deals with antigen presentation and signalling between B and T cells.

White, D. A. and Baxter, M. (1994) *Hormones and Metabolic Control*, 2nd edn, Edward Arnold.

> A comprehensive guide to the hormonal control of various aspects of human metabolism.

## Chapter 5

Kuffler, S. W., Nicholls, J. G. and Martin, A. R. (1984) *From Neuron to Brain*, 2nd edn, Sinauer Associates Inc.

> This book describes clearly and in considerable detail how nerve cells transmit signals, how these signals are put together and how this integration contributes to higher functions.

The Open University (1992) SD206 *Biology: Brain and Behaviour*, D. Robinson (ed.) The Open University.

## Chapter 6

Sato, S. (1990) *Advances in Neurology,* vol. 54. *Magnetoencephalography*, Raven Press.

> A collection of research papers on the use of MEG in diagnosis and its clinical role in relation to EEG.

Stewart, M. G. (1992) *Quantitative Methods in Neuroanatomy*, Wiley.

> Chapters 8 and 9 deal respectively with magnetic resonance imaging and biomagnetic imaging of the brain.

## Chapter 7

Glass, L. and Mackey, M. (1988) *From Clocks to Chaos*, Princeton University Press.

Goodwin, B. C. (1994) *How the Leopard Changed its Spots; the Evolution of Complexity*, Weidenfeld and Nicolson.

Lewin, R. (1992). *Complexity; Life at the Edge of Chaos*, Macmillan.

Winfree, A. T. (1987). *When Time Breaks Down; the Three-dimensional Dynamics of Electrical Waves and Cardiac Arrythmias*, Princeton University Press.

# *Acknowledgements*

We are very grateful to the assessors for this book, whose perceptive comments on early drafts were an important guide in preparing the final text: Professor Mike Berridge (Cambridge University), Dr David Katz (University College, London), Dr Peter Newell (Oxford University) and Professor Ron Pethig (University College of North Wales, Bangor).

We also gratefully acknowledge the help of Mike Bullivant in generating the chemical structures; Heather Davies for providing the electron micrographs; and Mary Cotterrell (University of Leeds) and Professor Miroslav Simic for their contributions during the early stages of preparing the book.

Grateful acknowledgement is made to the following sources for permission to reproduce material in this book:

### Chapter 1

*Figure 1.10* From Goltz, J. S. *et al.*, 'A role for microtubles in sorting endocytic vesicles in rat hepatocytes', *Proceedings of the National Academy of Sciences USA*, vol. 89, August 1992, pp. 7026–7030.

### Chapter 2

*Figure 2.10* Dr Barry Ganetsky, University of Wisconsin–Madison, Laboratory of Genetics.

### Chapter 3

*Figures 3.1, 3.27, 3.36, 3.37* From Stryer, L. (1988) *Biochemistry,* copyright © 1988 by Lubert Stryer, used with permission of W. H. Freeman and Co.; *Figures 3.3, 3.8, 3.14, 3.21, 3.24* Alberts, B. *et al.* (1983) *Molecular Biology of the Cell*, Garland Publishing, Inc., © 1983 by Bruce Alberts, Dennis Bray, Julian Lewis, Martin Raff, Keith Roberts and James D. Watson; *Figure 3.4* Adapted from illustration by Dana Burns in Bretscher, M. S. (1985) 'The molecules of the cell membrane', *Scientific American,* October 1985, copyright © 1985 by Scientific American, Inc., all rights reserved; *Figure 3.10* Biophoto Associates; *Figures 3.7a, 3.7b, 3.12* Mathews, C. K. and van Holde, K. E. (1990) *Biochemistry*, copyright © 1990 by the Benjamin/Cummings Publishing Company, Inc.; *Figure 3.13* Hyde, S. C. *et al.* (1990) 'Structural model of ATP-binding proteins associated with cystic fibrosis, multidrug resistance and bacterial support', reprinted with permission from *Nature*, **346**(6282), 26th July 1990, copyright 1990 Macmillan Magazines Ltd; *Figure 3.15* Professor D. Bainton (University of California); *Figures 3.18, 3.19, 3.20, 3.52, 3.57, 3.59, 3.61, 3.62* Courtesy of Professor M. Berridge; *Figure 3.22* Dr Nigel Unwin, MRC Laboratory of Molecular Biology, Cambridge; *Figures 3.35, 3.39* Linder, M. E. and Gilman, A. G. (1992) 'G proteins', *Scientific American,* July 1992, copyright © 1992 by Scientific American, Inc., all rights reserved; *Figure 3.51* Bezprozvanny, I., Watras, J. and Ehrlich, B. E. (1991) 'Bell-shaped calcium-response curves of $IP_3$ and calcium-gated channels from endoplasmic reticulum of cerebellum', reprinted with permission from *Nature*, **351**(6329), 27th June 1991, copyright 1991 Macmillan Magazines Ltd; *Figure 3.53a* From *Calcium and Cell Function:* vol. 1 Calmodulin; edited by Wai Yiu Cheung, Academic Press, New York, 1980; *Figure 3.53b* Courtesy of Dr Y. S. Babu and Dr W. J. Cook; *Figure 3.60* Berridge,

M. J. (1993) 'Inositol triphosphate and calcium signalling', Reprinted with permission from *Nature*, **361**(6410), 28th January 1993, copyright 1993 Macmillan Magazines Ltd.

*Table 3.1* Alberts, B. *et al*. (1983), *Molecular Biology of the Cell*, Garland Publishing, Inc., © 1983 by Bruce Alberts, Dennis Bray, Julian Lewis, Martin Raff, Keith Roberts and James D. Watson; *Table 3.5* Linder, M. E. and Gilman, A. G. (1992) 'G proteins', *Scientific American,* July 1992, copyright © 1992 by Scientific American, Inc., all rights reserved.

## Chapter 4

*Figure 4.3* Williams, C., Nishihara, M., Thalabard, J-C., Grosser, P. M., Hotchkiss, J. and Knobil, E. (1988) 'The hypothalamic GnRH pulse generator of the rhesus monkey', in *Abstracts of the Proceedings of the Society of Neuroscience*; *Figure 4.11* Dr W. van Ewijk, Erasmus University, Rotterdam; *Figure 4.15* 'Nitric oxide as a signal in blood vessels', in Knowles, R. G. and Moncada, S. (1992), *Trends in Biochemical Sciences*, **17**, pp. 399–402, International Union of Biochemistry and Molecular Biology, Elsevier Trends Journals; *Figure 4.17* Mathews, C. K. and van Holde, K. E. (1990), *Biochemistry*, copyright © 1990 by the Benjamin/ Cummings Publishing Company, Inc.; *Figure 4.19a* Alberts, B. *et al*. (1983), *Molecular Biology of the Cell*, Garland Publishing, Inc., © 1983 by Bruce Alberts, Dennis Bray, Julian Lewis, Martin Raff, Keith Roberts and James D. Watson; *Figure 4.19b* Courtesy of Dr Michael Stewart.

## Chapter 6

*Figure 6.2* Jaffe, L. F., Robinson, K. R. and Nuccitelli, R. (1974) 'Local cation entry and self-electrophoresis as an intracellular localization mechanism', in *Annals of the New York Academy of Sciences*, **238**, pp. 372–389, New York Academy of Sciences; *Figure 6.3* Photo courtesy of CTF Systems Inc., Port Coquitlam, BC, Canada; *Figure 6.4* Photo courtesy of R & D Center, Osaka Gas Co., Ltd., Japan and Dr N. Nakasato, Department of Neurosurgery, Tohoku University School of Medicine and Kohnan Hospital, Sendai, Japan; *Figure 6.5* Courtesy of Tony Bailey, Institute of Psychiatry, London; *Figure 6.6* Dr Richard B. Frankel, CalPoly, San Luis.

## Chapter 7

*Figure 7.2* Williams, C., Nishihara, M., Thalabard, J-C., Grosser, P. M., Hotchkiss, J. and Knobil, E. (1988) 'The hypothalamic GnRH pulse generator of the rhesus monkey', in *Abstracts of the Proceedings of the Society of Neuroscience*; *Figure 7.3* Filicori, M., Santoro, N., Merriam, G. R. and Crowley, W. F. Jr (1986) 'Characterization of the physiological pattern of episodic gonadotropin secretion throughout the human menstrual cycle', *Journal of Clinical Endocrinology and Metabolism*, **62**, pp. 1136–1144, Williams and Wilkins; *Figure 7.7* Njus, D. *et al*. (1981) 'Precision of the Gonyaulax clock', *Cell Biophysics*, **3**, p. 225, Humana Press, Inc.; *Figure 7.11* Courtesy of Patricia Lakin-Thomas, Department of Plant Sciences, Cambridge; *Figures 7.14, 7.15* Goldberger, A. L. and Rigney, D. R. (1988) 'Sudden death is not chaos', in Kelso, J. A. S., Mandell, A. J. and Schlesinger, M. F. (eds), *Dynamic Patterns in Complex Systems*, World Scientific Publishing Co. PTE Ltd; *Figure 7.16* Adapted from H. J. J. Wellens *et al*. (1972) 'Ventricular fibrillation occurring on arousal from sleep by auditory stimuli', *Circulation*, **46**, pp. 661–664, by permission of the American Heart Association, Inc. and the authors; *Figures 7.17, 7.18* Winfree, A. T. (1987), *When Time Breaks*

# Index

Entries in **bold** are key terms.